George Stewardson Brady

A monograph of the free and semi-parasitic Copepoda of the British islands

George Stewardson Brady

A monograph of the free and semi-parasitic Copepoda of the British islands

ISBN/EAN: 9783744726764

Printed in Europe, USA, Canada, Australia, Japan

Cover: Foto ©Andreas Hilbeck / pixelio.de

More available books at **www.hansebooks.com**

A MONOGRAPH

OF THE

FREE AND SEMI-PARASITIC

COPEPODA OF THE BRITISH ISLANDS.

BY

G. STEWARDSON BRADY, M.D., F.L.S.,

PROFESSOR OF NATURAL HISTORY IN THE UNIVERSITY OF DURHAM COLLEGE
OF PHYSICAL SCIENCE, NEWCASTLE-ON-TYNE; CORRESPONDING
MEMBER OF THE ZOOLOGICAL SOCIETY OF LONDON,
OF THE ACADEMY OF NATURAL SCIENCES
OF PHILADELPHIA, ETC.

VOL. III.

LONDON:
PRINTED FOR THE RAY SOCIETY.
———
MDCCCLXXX.

A MONOGRAPH

BRITISH

FREE AND SEMI-PARASITIC COPEPODA.

VOL. III.

GENERAL ANATOMY AND DEVELOPMENT
OF THE COPEPODA.

In preparing this monograph it was not part of my plan to enter at all into the consideration of the physiology or internal anatomy of the Copepoda, but, in compliance with the wishes of some of my friends, expressed while the first volume was passing through the press, I have put together, in the form of a preface to this third volume, a condensed account of some of the more important observations which have hitherto been made on this part of the subject. It will be at once apparent that what I have attempted is nothing more than a general outline, in the production of which I have been greatly indebted to the works of Claus, Gegenbaur, and Huxley. Had I included the truly parasitic species, in whose anatomy, physiology, and general habits of life so many points of the highest interest occur, I should have been travelling beyond

VOL. III.   A

the professed limits of my work, and have extended
this preface beyond reasonable bounds,—on which
account those species are noticed in the most cursory
manner, and only in elucidation of the proper subject
of the memoir.

The non-parasitic COPEPODA may be described as
ENTOMOSTRACA, having an elongated body of variable
form, but generally cylindrical, without a bivalved
shell, and showing more or less completely a division
of the body into numerous rings, or somites. There
are two pairs of antennæ, three pairs of prehensile
and masticatory, or suctional, mouth organs, and five
pairs of feet, adapted chiefly for swimming. The
females are mostly fertilized by means of spermatophores; the ova are usually carried in external ovisacs;
when first hatched the larvæ have only three pairs of
limbs, and go through several metamorphoses before
attaining the mature form.   .

The parasitic species, at one end of the series
approach very nearly in structure and general appearance to the non-parasitic; at the other end they are
extremely different, exhibiting, especially in the
males, many most remarkable examples of retrograde
development, so that without the study of their
metamorphoses it would be quite impossible to re-
cognise them as Copepoda, or even as Crustacea of
any kind. Yet even in these degraded forms—at any
rate in the females—natatory limbs in a very much
atrophied condition are almost constantly found.

GENERAL FORM.—The animal is usually somewhat
pear-shaped, rounded in front and tapering towards the

hinder extremity, convex on the dorsal and flattened on the ventral surface. The degree of dorsal convexity, however, is variable, the most common form of the body being sub-cylindrical, but in many cases—notably amongst the Pœcilostoma, Siphonostoma, and in some genera of the Harpacticidæ (*Zaus, Peltidium, Porcellidium, Idya, Scutellidium*)—the animal is markedly flattened, constituting a type of structure almost as widely distinct from the normal Copepoda as are the Isopoda from the Amphipoda, amongst the sessile-eyed Crustacea. But in the genera of Harpacticidæ here referred to, the flattened form does not coincide with any deep-seated differences of structure, for, indeed, in some normal genera (*Thalestris, Laophonte*) we find certain species assuming this form, but retaining all the other characteristics of the genus to which they belong. The directly opposite type of structure, in which the animal is compressed laterally, occurs only in the curious genus *Amymone*, a group of rare occurrence so far as our present knowledge extends, not having been noticed outside of the European area.

The front of the body is covered by a membranous shield or carapace, which falls rather loosely over the bases of the limbs below, but behind is continuous with the coverings of the thoracic somites; in front it is usually projected from the forehead in the middle line, forming a rostrum, short or entirely absent in some few cases, but often moderately long and curved. In the Calanidæ it is usually long and slender, much curved, and furcate nearly to its base. Behind the rostrum is commonly placed the median eye, often single to all appearance, but really composed of two closely approximated lateral eyes, embedded in a mass of black or crimson pigment.

The head and the first thoracic segment are usually fused together, a fact which may be recognised by the position of the first pair of swimming feet (the first thoracic appendages), these limbs being generally fixed to the hinder part of the first body-segment, which is thus seen to be composed of all the cephalic, and the first of the thoracic, somites. In some cases a transverse indentation may be noticed, probably a trace of the " cervical suture " which is so conspicuous a feature of the carapace in Crayfishes and Lobsters. Instances are frequent, however, in which the head is quite distinct from the thorax.

Theoretically, the Copepoda, like other Crustacea, are composed of twenty or twenty-one somites (twenty-one according to most authors, but twenty if we follow Huxley, who does not look upon the telson or last abdominal segment as a true somite), the entire series not being developed, however, in any one animal. Sometimes one or many of the somites are suppressed, at other times several are united into one segment, the real nature of which is rendered evident by the attachment to it of several appendages, each pair indicating the position of an anchylosed somite. Probably in no case amongst the Copepoda can more than sixteen or seventeen somites be recognised by enumeration either of appendages or segments. The somitic appendages or limbs, though very various in form and function, may all be reduced to three component parts—a basal portion or *protopodite*, which gives support to two branches, termed respectively *endopodite* and *exopodite*. These parts are most clearly developed in the swimming-feet, which are distinctly made up of a peduncle and two branches, but in the mouth organs the same structure may be traced, though often modified to such an extent as to be obscure and difficult of recognition.

The cephalon in the Copepoda is composed of six somites, its appendages being one pair of eyes, two pairs of antennæ, one pair of mandibles, one pair of maxillæ, and two pairs of foot-jaws; the *thorax* consists of five somites, and has five pairs of appendages in the form of swimming-feet; the *abdomen* has no appendages, but consists of five somites and terminates in a forked tail, which ought probably to be considered as a sixth somite; in the female the first two abdominal somites are generally united, forming one large genital segment with a pair of vulvar apertures. In parasitic species the abdomen is often very much reduced in size, both as respects the number and bulk of its somites, and this is the case also in some genera which are only partially parasitic, as *Corycæus, Acontiophorus, Artotrogus* and others. As regards the cephalic appendages, it must be noticed that the so-called two pairs of foot-jaws are in reality portions of one and the same somitic appendage, but having the appearance of perfectly distinct organs they have come to be con-

sidered as such, and their characters so taken note of
by all systematists. The process of development,
however, has been traced by Claus, so as to leave no
doubt as to the true nature of the organs.

APPENDAGES OF THE HEAD.—*The eyes* in their simplest
form—in *Cyclops,* for instance—appear as a red or
black spot in the middle of the frontal region, directly
over the brain, with which they are connected by a large
nerve, the spot when closely examined being found
to consist of two lateral eyes, closely approximated
and embedded in a mass of pigment; the visual part
of the apparatus is composed of two refracting bodies,
or crystalline cones, and when more highly developed
may possess numerous lenses, so as to form something
like a facetted cornea. In some cases the eyes are
widely separated, and have between them, in the
median line, a simple, globular, pedunculated eye
(*Pontellinæ*); in other cases, as in some *Corycæidæ,*
the median eye is very small, while the lateral eyes
are large, destitute of pigment, and consist of simple
highly refracting lenses. In some parasitic species
only are the eyes entirely wanting. In *Pleuromma*
there is a supplementary eye, consisting of lenses
with black pigment matter, on one of the thoracic
segments.

The anterior antennæ are usually large and conspicuous organs, rising from hollows in the front of the
head on each side of the rostrum. They act in many
species (*Calanidæ, Cyclopidæ*) as very powerful swimming organs, it being by their agency chiefly that these
animals propel themselves through the water; the

motion is thus a succession of rapidly recurring jerks corresponding with the separate sweeps of the antennæ. These organs vary greatly in length and in the character of their setose armature. In the *Harpacticidæ*, as well as in most parasitic and semi-parasitic species, the length of the anterior antenna falls short of, or at any rate does not usually much exceed, that of the first division of the body (carapace), while in the *Cyclopidæ* it often equals, or even exceeds, that of the cephalothorax. In many *Calanidæ* the length of the antennæ is still greater, not unfrequently exceeding—sometimes very much exceeding—that of the entire animal. There is, however, no instance of this kind among British species, nor, so far as I know, among any but distinctly pelagic forms. And it is remarkable that, with this extreme length of the antennæ, there is usually asssociated a greatly increased development of the apical lash of swimming-setæ with which the organ is armed. Sometimes also, as, notably, in the pelagic genus *Euchæta*, the marginal hairs, though few, become wonderfully long. The antennæ have, however, other important functions besides those of locomotion; some of the variously formed setæ with which they are in most cases largely provided act, no doubt, as organs of special sensation— the more simple hairs, perhaps, as tactile, the flattened and club-shaped setæ as olfactory organs. The flattened, ensiform kind of appendage is seen on the antennæ of most of the *Harpacticidæ* (*e. g. Idya furcata*, Plate LXVII, fig. 2 *a*), and the club-shaped form is seen in great force in *Isias clavipes* (Plate VII, figs. 4, 5) as

well as in other *Calanidæ;* these last-named organs are
always most largely developed in the males, and it is
not unlikely that they are subsidiary sexual organs, possibly endowed with a highly-developed sensuous faculty.
Besides the forms of antennal appendages here noted
other modifications exist, of which the special uses are
at present unknown.

The anterior antennæ serve yet another essential
purpose, being adapted in the males as clasping organs;
they, together with the fifth pair of feet, are in very
many species specially fashioned so as to insure a firm
grasp of the female. The number of joints found in the
anterior antennæ varies from five or six in some semi-
parasitic species (*Lichomolgus*) and eight or nine in the
Harpacticidæ, to twenty-four or twenty-five in the
Calanidæ. In the females the antennæ are always
alike on both sides of the body, but in the males
of many of the *Calanidæ* the right antenna is modified
for the purpose of clasping; while in all other non-
parasitic species the male antennæ of both sides are
specialized for that purpose. In the semi-parasitic
Corycœidæ, Saphirinidæ, &c., there is little or no
sexual difference in the anterior antennæ, the clasping
function devolving on the posterior pair. The par-
ticular structural adaptations differ in different cases;
in the *Calanidæ,* as before stated, the right antenna
only is differentiated, the alteration consisting in one of
the joints not far from the apex being so articulated
as to form a hinge, by means of which the distal portion
of the limb can be flexed upon the basal portion.
Above and below the hinge the inner margins of the

antennæ are frequently armed with denticulated plates, giving a firmer grasp. A very remarkable instance of this structure exists in *Candace pectinata* (Plate X, fig. 2), also in the genera *Centropages, Pontella, Parapontella,* &c.; other genera, such as *Temora* and *Diaptomus* (Plate VI, fig. 7), are provided with one or more strong spines in lieu of, or in addition to, the denticulated plates. These spines are situated at various points of the internal surface of the antenna above the hinge-joint, and the limb itself is more or less swollen in the same situation to give room for a powerful flexor muscle. In the *Pontellinæ* this enlargement of the limb is excessive (Plate X A, fig. 2; Plate XI, fig. 2). In the males of many *Calanidæ*, however, there is little or no difference of form between the antennæ of the two sides; the difference is very slight also in the *Misophriidæ*. In most *Cyclopidæ* there are found differences of a kind similar to those already described, but affecting equally both right and left antennæ (Plate XVII, fig. 5; Plate XXII, fig. 16); there are here no denticulated plates and few spines, but the limb is distinctly geniculated near the base as well as near the apex, and the articulations of the terminal segments are likewise very mobile. Amongst the *Notodelphyidæ* the structure of the male antennæ is somewhat similar to that of the *Cyclopidæ*, but by no means so well marked. In the *Harpacticidæ*, where the anterior antennæ are very much smaller in comparison with the size of the animal, the hingement of the male organ is not quite so obvious, though still sometimes quite of Cyclopid type (*Canthocamptus minutus*, Plate XLIV, fig. 3; *Robertsonia tenuis*, Plate XLI, fig. 3); the apex is, however, often strongly clawed (*Longipedia*, Plate XXXIV, fig. 2; *Euterpe*, Plate XL, fig. 2; *Tachidius*, Plate XXXVII, fig. 3, &c.), and very frequently the two or three joints at the proximal side of the claw are coalescent and greatly enlarged, forming a pyriform or subglobose swelling for the reception of strong muscular bands; examples of this structure are seen in the genera *Tachidius* (Plate XXXVII, fig. 3), *Harpacticus* (Plate LXIV, figs. 2, 13), *Jonesiella* (Plate XLVIII, fig. 3), and in many others. The spinous and setose armature of the antennæ is, as a general rule, more profuse in the males of all families; a good illustration of this character is seen in the case of *Longipedia* (Plate XXXIV, figs. 2, 3). In *Pœcilostoma* and *Siphonostoma* the sexual distinctions of the anterior antennæ are not strongly marked, consisting chiefly of imperfect hingements or contractions of the articulating surfaces between various joints of the male organ.

The posterior or second pair of antennæ are generally much smaller than the anterior pair, and consist in most cases of two branches; they seem to be used both as swimming and prehensile organs, and in parasitic and semi-parasitic species are specialized for prehension much more decidedly than the anterior pair. The main branch consists usually of three or four joints, to the basal or second joint of which is attached a "secondary" or "inner" branch of smaller size, and composed of one or several joints. This branch is in some cases altogether absent; in others it is equal in size to the primary branch, and may be numerously jointed (*Longipedia*, *Calanidæ*, *Misophriidæ*). In *Cyclopidæ* the secondary branch is wanting, and in most of the *Harpacticidæ* it is reduced to small dimensions or, in some few cases, is quite absent. The limb is generally curvate, dilated at the apex, and provided with numerous curved or geniculated setæ at the extremities and over the margins of both branches. In the *Pæcilostoma* and *Siphonostoma* it consists of only one branch, which in the male of the *Corycæidæ* (Plate LXXXIII, figs. 13, 14) is powerfully clawed, and is used chiefly as a prehensile and clasping organ.

The mandible consists, in its fully developed form, of a masticating portion and a " palp;" the first-named division is in the form of an elongated, more or less triangular plate, dilated at the distal extremity and cut up into a variable number of tooth-like processes, these being sometimes only slight serrations, sometimes large and powerful. The palp is variously formed; in the *Calanidæ*, most of the *Harpacticidæ*, and in many *Cyclopidæ*, it is composed usually of a large basal joint, from which spring one or two small setiferous branches; these branches are generally 1- or 2-jointed, but may be absent altogether. In the sub-families *Porcellidiinæ* and *Idyinæ* the setiferous portions of the organ are very largely developed, forming large laminated appendages, which are fringed with densely ciliated filaments (Plate LXVIII, fig. 4; Plate LXIX, fig. 11). In the genus *Cyclops*, while the biting part of the mandible is well developed,

the palp is atrophied, and is represented only
by two or three ciliated setæ. In the *Corycæidæ*
the mandible is small and weak, its palp obsolete
or reduced to very small dimensions, while in some
Siphonostoma* the mandible itself is converted into
a long and slender piercing style, which is enclosed
in a tube resembling considerably the antlia of Lepi-
dopterous insects, and composed of prolongations
of the upper and lower lip (Plate XCIII, fig. 3);
in *Cyclopicera*, however (a genus here included
amongst Siphonostoma), there is an intermediate
condition of things; the mandible is very much elongated, slender, and finely toothed at the apex, being,
in fact, almost stylet-shaped, but not enclosed in any
sheath; the palp, also, is quite rudimentary (Plate
LXXXIX, fig. 4).

The maxillæ are small appendages, composed of a
chewing portion, which is divided at the apex into
numerous rather long and slender curved teeth, and
of a complex, lobed, and setiferous palp, which
frequently has filamentous appendages, possibly of a
branchial character. In *Cyclops* the maxillæ are very
small in all their parts; in the *Calanidæ*, on the
contrary, they are largely developed and possess
numerous plumose filaments (Plate I, fig. 5); in the
Notodelphyidæ, also, the filamentous appendages are
large (*e. g.* Plate XXIX, fig. 5). The *Harpacticidæ*

---

* The homologies of the mouth-organs in this group have been dis-
cussed by Claus, Thorell, Buchholz, and others. The subject is not
without difficulty, and is treated under the description of the *Siphonostoma* (p. 26) at greater length than would be suitable in this prefatory
notice.

have the maxillæ very small, but numerously divided; there is usually a chewing portion pretty strongly toothed, and a palp which is divided into two or three setiferous digits, and has likewise two lateral (? branchial) offshoots (Plate LVII, fig. 2, Plate LX, fig. 6, Plate LXII, fig. 5, &c.), but this arrangement is subject to endless variation as to the number and development of the various parts. In parasitic species the maxillæ, though usually present, are often atrophied, and in the semi-parasitic forms belonging to the groups *Pœcilostoma* and *Siphonostoma*, are extremely minute, and sometimes attached to the base of the mandibles.

The anterior and posterior foot-jaws (which are in reality only the exo- and endo-podites of a single somitic appendage) do not present features requiring any lengthened description. They are in all cases adapted for prehension. In the *Calanidæ, Cyclopidæ,* and *Notodelphyidæ,* the two pairs are not very dissimilar in structure, consisting generally of from four to six joints, which are in most cases marginally produced into digitiform or wart-like eminences on their inner side, and are more or less densely setiferous. In the *Harpacticidæ,* while the first foot-jaw is like those of the preceding families, the second is usually in the form of a strongly-clawed hand, and in some subfamilies (*Idyinæ,* &c.) both pairs assume this character; such, also, is the case in the semi-parasitic *Pœcilostoma* and *Siphonostoma.* The subfamily *Longipediinæ,* unlike the rest of the *Harpacticidæ,* has the posterior foot-jaw destitute of a clawed hand, those organs in *Ectinosoma* being excessively slender, and almost linear in form.

THE APPENDAGES OF THE THORAX are, in their simplest form, five pairs of swimming-feet, each foot consisting of a 2-jointed base and two 3-jointed branches, but the number of joints, though never exceeding three, is within that limit, subject to a good deal of variation. The second, third, and fourth pairs are in almost all cases constructed as simple swimming-feet, without any other function, but the first pair, not unfrequently (as in most *Harpacticidœ*) has one of its branches converted into a clawed prehensile limb, and the fifth pair shows very extensive modification in almost every case, often differing very considerably in the two sexes, and sometimes being very much reduced in size, or even altogether wanting. Though, as has been already said, the second, third, and fourth pairs of feet are constructed simply as swimming organs, and present no peculiarities of form, there are some exceptions to that statement. In *Metridia armata* the inner branch of the second foot in the male has a very remarkably excavated notch with spinous margin (Pl. LVI, fig. 20 *b*), and some foreign species, belonging, probably, to two distinct genera, *Metridia* and *Undina*, exhibit a similar structure,* sometimes in both sexes. Again, in the genus *Harpacticus* the inner branch of the *second* foot of the male has the middle joint produced downwards into a strong spine, which varies in character according to species, but, in some

---

* These species will be described in the ' Report on the Copepoda ' taken during the voyage of H.M.S. " Challenger."

shape or other, is always present (Pl. LXIV, figs. 7, 16; Pl. LXV, fig. 11). The third foot of the male in the same genus is sometimes (perhaps not always) converted into a clasping organ, the outer branch being bent across the inner and having its last joint armed with several strong spines (Pl. LXIV, fig. 20). In the *Calanidæ* and *Misophriidæ* the fifth pair of the male is usually specially adapted as a clasping organ, the limb of one or both sides being reduced to a single branch, and provided with an armature of spines or hooks, which either entirely supersedes the swimming function, as in *Temora*, or is superadded, as in *Centropages;* but in other genera belonging to these families the sexual alteration of the limb is not very great (*Calanus*). Some pelagic genera, which are not represented in the British seas (*Euchæta, Undina*), though possessing in the male a strongly developed prehensile fifth foot, have in the female only four pairs of simple swimming-feet. In *Undina* the male fifth foot is remarkably long and very fantastic in shape, reaching sometimes even beyond the extremity of the caudal segments. The organ may not unfrequently be seen with spermatophores adherent to its apex,* and is possibly used as the means of conveying these bodies to the vulva of the female.† It is remarkable, too, that in species so constituted (especially in *Undina Darwinii*, Lubbock) the spermatophores are very commonly found affixed in a futile manner to the back of the thoracic rings of

* A similar condition is figured by Dr. Claus in the case of *Euchæta prestandreæ* (' Die frei lebenden Copepoden,' pl. xxx, fig. 9).

† A similar function is performed by the maxillary palps (chelæ) of male spiders and by the hectocotylised arms of some cuttle-fishes.

the female. In the *Cyclopidæ* and *Notodelphyidæ* the fifth feet are usually rudimentary and alike in both sexes, and in the *Harpacticidæ* they take the form of small, marginally setose, foliated expansions, slightly different in the two sexes, but generally larger in the female, in which sex they serve sometimes as a covering and support for the external ovisacs. In the semi-parasitic groups these organs are generally small, 1- or 2-jointed, and alike in both sexes.

THE NERVOUS SYSTEM of the Copepoda is described as consisting of a brain, which gives off various sensory nerves, a sub-œsophageal ganglion, and a ventral nerve cord, on which are situated ganglionic enlargements; the antennary nerves are also thickened, forming ganglionic rings.

From an investigation of the nervous system of *Cyclops*, Mr. Marcus M. Hartog, F.L.S., of Owen's College, Manchester, has recently made out some other points which he kindly allows me to insert here. The most important of these are, that " ganglionic swellings are found near the terminations of all sensory nerve fibres; that the ventral nerve cord gives off at the end of the third segment of the body a pair of superficial cutaneous nerves, and at the fourth segment two pairs, one to the rudimentary fifth legs and another to two ventral muscles which rise from the sternal portion of the fifth segment. In the first abdominal segment is the fork described by Claus and Leydig, but this takes its origin from the superficial (ventral) aspect of the cord which is continued onwards under the colleterial gland. After running obliquely outwards each branch of the fork subdivides into two, an anterior and a posterior branch, both sensory. At the commencement of the third abdominal segment the ventral cord forks, its branches diverge slightly in this segment, but more in the next, rising to the sides of the intestine, and having the ventral muscles of this segment superficial to them. In the last segment they have left the intestine and run about the horizontal median plane straight into the axis of either branch of the furca.

" The ventral cord in Cyclops is not differentiated into distinct ganglia up to the second free (third) thoracic segment : beyond this is an enlargement containing ganglion cells at the posterior end of the fourth, and another (very small) in the last thoracic segment."

THE ORGANS OF SENSE, so far as they exist in the tactile and olfactory rods of the antennæ, have already received brief notice. Mr. Hartog has recently described certain vesicles in the frontal region of Cyclops, and others attached to the bases of the fifth pair of feet and seated on a ganglionic enlargement of the nerve supplying the feet, which vesicles he believes to be auditory organs. In the male they contain one or more highly refracting bodies floating freely in the interior. Claus found a pair of these vesicles in the brain of *Calanella*, and has figured them in ' Die frei lebenden Copepoden ' (plate vii, fig. 9).

THE DIGESTIVE CANAL has a short, straight gullet, a large stomach, often with two cæcal tubes, and an intestine opening on the dorsal aspect of the last (or last but one ?) abdominal segment. The hinder portion of the alimentary canal is perhaps also uriniferous, but there exist near the bases of the foot-jaws other glandular organs, which may have a renal function. In the males of parasitic Copepoda the digestive canal disappears entirely.

RESPIRATION.—If we are right in assigning to certain appendages of the mouth-apparatus (to which reference has been already made; see pp. 10, 11*) a branchial function, then we cannot altogether assent to the commonly - received belief that respiration in the Copepoda is carried on entirely by the dermic and intestinal surfaces of the body, without the intervention of any specialized respiratory apparatus. A sub-rhythmic contraction of the hinder extremity of the gut has, however, been noticed by Mr. Hartog in several Copepoda (*Cyclops, Diaptomus, Canthocamptus*), and by several observers in other Crustacea (*e. g. Astacus, Limnadia, Daphnia*): this is, no doubt, a respiratory movement.

CIRCULATION.—In many Copepoda no special circulatory organs have been found; but in some there is a tubular heart, situated in the last thoracic segment, which drives forward the blood by a short vessel to the brain and anterior parts of the body, the blood

* There can be no doubt that these setiferous plates are homologous with the structures called in the Ostracoda *branchial laminæ* by G. O. Sars. But they do not appear to have any special internal circulation of the vital fluids, and if their function be branchial, they must act only by propelling waves of aerated water over the neighbouring surfaces. It is, perhaps, on the whole, most probable that the currents produced by these ciliated appendages are subsidiary to nutrition rather than to respiration.

then passing through lacunæ scattered throughout the tissues of the animal, and finding its way back to the heart, which it enters by slits in the walls of that organ.

REPRODUCTION. — The sexes in the Copepoda are always separate, sexual differences showing themselves even externally in the form and structure of the body; in some, especially in parasitic species, the dimorphism is most remarkable, the male becoming little more than a motionless sperm-sac attached to the body of the female; but in the species which come within the scope of this memoir the males are usually smaller, more active, and less numerous than the females, the chief external distinctions being found in the almost constant conversion of the anterior antennæ—less constantly of the fifth pair of feet, and occasionally also of the posterior antennæ and foot-jaws—into clasping organs. The ovaries and testes are placed in the middle or in the sides of the cephalothorax, communicate with accessory glands, and have efferent canals, which open by distinct apertures on the sides of the first (or conjoined first and second) abdominal somites. The efferent canal of the ovary may be simple, or may give off laterally a number of pouches, which hold the eggs (*Corycæidæ*), while in some parasitic species it forms several terminal coils, in which the eggs are detained; in the *Notodelphyidæ* the duct is converted into a large dorsal pouch or pseudo-uterus, covered only by the integument, in which organs the ova undergo partial development. In the free Copepoda, however, the ova pass at once into two (often coalescent) external ovisacs attached to the first abdominal segment. the coating of the ovisac being formed by the secretion of a special gland, situated near the termination of the efferent duct, an enlargement of which forms in many cases a "receptaculum seminis." In the males the free-living forms have a simple testis; many of the parasitic and semi-parasitic (*Corycæidæ*, *Sapphirinidæ*) a double testis, with two distinct *vasa deferentia*, the right duct being sometimes atrophied. In the coiled portion of the duct are formed the spermatophores—masses of spermatozoids enclosed in a capsule of hardened mucus, and usually fusiform or club-shaped. During copulation the male affixes one or more of these bodies near the vulvar aperture of the female, the contents passing into the *receptaculum seminis,* and fertilizing the ova either in the interior of the body or during their passage into the ovisacs. In some cases the seminal fluid appears to be inserted directly into the vulva without the intervention of a spermatophore.

THE DEVELOPMENT of the free Copepoda from the moment of rupture of the ovum to the attainment of matured form presents a complex series of metamorphoses. The parasitic species present some of the best marked examples of "retrograde development" to be found in the whole animal kingdom, but these do not come within the limits of our present subject. The form of the young Copepod on its escape from the egg is that known as *Nauplius,* having been described by Müller under that name before its relation to the Copepoda was known. The larva in this stage is oval, has a single frontal eye, three pairs of limbs arranged round the mouth, and no frontal appendages; the mouth-organs proper are entirely absent, and the posterior part of the body has no appendages except a couple of setæ in the neighbourhood of the anus. The anterior portion of the body is equivalent to the three anterior cephalic somites, its three pairs of limbs becoming eventually antennæ and mandibles. At the first moult the body becomes elongated and new limbs appear in the following order :—a fourth and fifth pair representing respectively the maxillæ and foot-jaws; a sixth and seventh which become the two anterior pairs of swimming-feet. At this stage the larva still resembles a *Nauplius,* and does not take on a Cyclopoid appearance until after the next moult.* It then resembles more closely the adult *Cyclops* as to the antennæ and mouth-organs, but the number of limbs and somites is smaller ; the body in this condition is composed of an oval cephalothorax, three thoracic and one long terminal segment, which in succeeding moults becomes forked. In the *Cyclopidæ* the posterior antennæ and the mandibles lose their accessory branches, but in other families these parts are usually retained. All the free, and many of the parasitic species pass through a further series of moults, in the course of which the still-wanting limbs and body-segments appear, the limbs attaining, by successive

* In some Nauplii, if not in all, the terminal part of the intestine is subglobular, and contracts periodically like the "contractile vesicle" of a Rotifer.

steps, their full number of joints and perfect development.*   Those parasites which miss the *Nauplius*
stage are hatched in *Cyclops*-form; many of the
retrogressive species become fixed to some animal,
segmentation is lost, limbs and eyes disappear or
become atrophied; the males, often dwarfed, being
permanently fixed near the sexual apertures of the
female.

* For details of this process in the genus *Cyclops*, see vol. i, p. 100.

NOTE.

Change of Generic Names.

The generic name *Lophophorus* (vol. i, p. 121), having been previously used to designate a genus of Phasianidæ, must be withdrawn. I therefore propose to substitute the word *Pterinopsyllus* (πτέρινος—ψύλλος).

Cylindrosoma (in Table of Genera, vol. i, p. 31) is for a like reason discarded for *Cylindropsyllus*, and *Solenostoma* (*loc. cit.*) for *Acontiophorus*.

BRITISH COPEPODA.

Section II.—Pœcilostoma, *Thorell*.*

Thorell's division of the Copepoda into three groups, Gnathostoma, Pœcilostoma, and Siphonostoma, the distinctions between which are found in the characters of the mouth-organs, is disapproved by Claus and some other authors, chiefly, as I understand them, on the ground of the gradual lapse of one series into the other rendering it impossible to draw perfect lines of demarcation, but partly, also, on the ground of a difference of interpretation of the homologies of some of the appendages. While differing from M. Thorell as to the nature of some of these organs, I myself think that his proposed division is a very natural one, the three groups presenting characters which, though differing in degree in various species, do point, on the whole, to habits of life very remarkably different, and deserving of expression in any natural classification.

The three groups are defined by M. Thorell as follows :

Series 1. Gnathostoma.

Os mandibulis duabus libens tribusque paribus maxillarum instructum, siphone nullo.

* See p. 31, vol. i.

Series 2. PŒCILOSTOMA.

Os mandibulis et siphone carens, maxillarum paribus
3—1 (—0) instructum.

Series 3. SIPHONOSTOMA.

Os in siphonem, mandibulas 2 plerumque inclu-
dentem, productum, et maxillarum paribus 3—0 instructum.

As regards the debateable anatomical points, it may
be useful if I quote, in the first place, some remarks
of M. Thorell, taken from a letter which he was good
enough to address to me some few years ago—before
the publication of Dr. Claus's 'Neue Beiträge zur
Kenntniss parasitischer Copepoden.' M. Thorell writes
as follows :—" You know, of course, that I have proposed to divide the Copepoda into three parallel series,
Gnathostoma, Pœcilostoma, and Siphonostoma, and
that I consider the Pœcilostoma to be characterized
by having the parts of the mouth *free*, and formed
for *stinging* or *licking*, as also by the *mandible being
absent*. This view of the oral apparatus of the Pæcilostoma has been accepted by Claparède and a few
others, but it is not admitted by, for instance, Claus,
who considers that the Pœcilostoma have true mandibles, and who rejects the subdivision proposed by me.
The reasons which induced me to believe that the
Pæcilostoma were destitute of mandibles, and that
what Claus calls mandibles are the true maxillæ, and
that his maxillæ are maxillar-palpi, were—*first*, that
the "mandibles" are always placed far more backwards than the true mandibles of the Gnathostoma;

secondly, that I sometimes, as, for instance, in *Licho-molgus albens,* found a small longitudinal, semi-pipe-formed depression or groove exactly at the place where the sipho of, for instance, *Dyspontius* and *Ascomyzon* is inserted, and which I therefore considered to indicate the place where the sipho and mandibles ought to be found if any mandibles existed; and *thirdly* and chiefly, that sometimes, as in the genus *Lichomolgus,* the so-called ' maxillæ ' are fixed on the ' mandibles ' (quite as an ordinary palpus is fixed on a mandible or a maxilla), and directed *from* the oral aperture, a cir-cumstance with which I could find nothing analogous in the class Crustacea, supposing Claus' ' mandibles ' really to be mandibles."

The greatest difficulty which besets the discussion of this question is the minuteness of the mouth-organs in these animals, and the liability to displacement or mutilation of the various parts in conducting a dissec-tion, so that the organs of one and the same species will often present very different appearances in differ-ent preparations of the animal. There can be no doubt, however, that the fact so strongly insisted on by M. Thorell,—that of the coalescence, in *Lichomolgus,* of the maxilla and mandible (or maxilla and palp)—does really exist: the question remains, What is this palp-like organ? In appearance it is not unlike the poorly-developed mandible- or maxilla-palp of many Gnathos-toma, but it is also much like a single branch of such a maxilla as we find in the genus *Cyclopicera* or *Arto-trogus,* so that not much can be learned by comparison of structure only. The point next to be considered is the nature of the appendages,—by Thorell called the first pair of maxillæ, by Claus the mandibles,—to which the palp-like organ is attached.

In considering this question we shall do well to take a somewhat wider survey than merely of the order Copepoda. Among the nearly related order Ostracoda, for the most part consisting of true Gnathostomous crustacea, we find a group,—including chiefly the genus *Paradoxostoma*,—in which the mouth is modified for suctorial purposes in a manner at once reminding one of the siphonostomous Copepoda. In *Paradoxostoma* the tubular mouth is formed by the coalescence of the labrum and labium, and the mandible assumes the form of a stilet, having a very slender filiform palp, the almost exact counterpart of the same organs in *Acontiophorus*, *Cyclopicera*, &c. (see Pl. LXXXIX, fig. 4, and Pl. XC, fig. 4). There can be no doubt, I think, that in the well-marked siphonostomous Copepoda, such as *Acontiophorus*, *Dyspontius*, and *Artotrogus* (*Ascomyzon*), the tubular mouth is formed, as in *Paradoxostoma*, by the union of the upper and lower lips, and that the filiform organs lying immediately by the side of the siphon (see Pl. XC, fig. 1 *c*, and Pl. XCI, fig. 6 *c c*) are modified mandibles and palps; in *Artotrogus*, indeed (Pl. XCIII, fig. 3 *b b'*), we find this stilet-shaped mandible distinctly toothed at its apex. In the genus *Cyclopicera*, of which I have fortunately collected and examined many specimens with great care, all the mouth-organs are largely developed; there is an unmistakable mandible with a well-developed palp, a distinct

2-branched maxilla, and a stout proboscidiform suctorial mouth (Pl. LXXXIX, figs. 4, 5, 6).   And it can
scarcely be doubted that the two-branched organ
shown in Plate LXXXVII, fig. 10 (*Acontiophorus
armatus*), is homologous with the mandible and palp
of *Cyclopicera nigripes* represented in·Plate LXXXIX,
fig. 4, and of *C. gracilicauda* (Plate LXXXIII, fig. 3).

To recur to the genus *Lichomolgus*.   If we examine
again the disputed organ (Pl. LXXXV, figs. 4, 12,
Pl. LXXXVII, fig. 3, Pl. LXXXVIII, fig. 11) we shall
find that it bears a very strong structural resemblance
to those appendages of siphonostomous genera which
have just occupied our attention, and the mandibular
nature of which is, I think, pretty conclusively shown.
I have therefore little doubt that this organ in *Lichomolgus* ought to pass for a mandible, and inasmuch as
the mandibular palp amongst the Siphonostoma is
sometimes nearly or quite suppressed, but the maxilla
never, and, moreover, as when the mandible-palp does
exist it assumes a form totally different from the palp-
like appendage of the mandible of *Lichomolgus ;* for
these reasons I am disposed to regard this appendage
as a rudimentary maxilla.   The opinion is confirmed
by a comparison of the maxilla of *Corycœus*, which is
connected with the mandible in a manner very like
that of the supposed maxilla of *Lichomolgus* to its
mandible (see Plate LXXXIV, fig. 10).

[*Genus* CYLINDROPSYLLUS, *nov. gen.*

Animal cylindrical, much elongated; head united
with the thorax and having a sharp rostrum; abdomen
4-jointed, as wide as the thorax, and not distinctly
separated from it.   Anterior antennæ short, 5-jointed;
posterior 2-jointed, destitute of a secondary branch.
Posterior (?) foot-jaw small, provided with an apical
curved spine and several marginal setæ (rest of the
mouth-organs unknown).   First four pairs of swimming-feet having the outer branch 3-, the inner 2-
jointed; branches of the first pair very short and
nearly equal; inner branch of the second, third,
and fourth pairs very short; fifth pair rudimentary,
1-jointed, foliaceous.

In the Table of Classification (vol. i, p. 31) this
genus is given as *Cylindrosoma;* which name, however,
being already in use, is here altered to *Cylindropsyllus.*
Though given in the table amongst Harpacticidæ, it
more probably belongs to the Pœcilostoma.   For the
present I content myself simply with a description of
the species, without attempting to assign it to any
recognised family.

1. CYLINDROPSYLLUS LÆVIS, *nov. sp.*   Pl. LXXXIV, figs.
1—8.

Animal of equal width throughout, the abdomen
being as wide as the cephalothorax and not separated
from it by any constriction; first body-segments as

long as the two next following; rostrum as long as the first joint of the anterior antenna; abdomen equal in length to the cephalothorax, and composed of four nearly equal joints. Anterior antenna (fig. 2) about as long as the first segment of the body, 5-jointed, joints nearly equal in length, except the fourth, which is much shorter than the rest; posterior antenna 2-jointed, the last joint bearing about eight apical setæ. First pair of feet short; second and third (fig. 5) longer; fourth (fig. 6) rather longer than the third; the external branches of the second and third pairs bear long and slender marginal spines, but those of the fourth are very weak and small. The fifth pair (fig. 7) consists of one small subquadrate joint, which is fringed distally with five or six rather long setæ. The caudal segments are very small, about twice as long as broad, but scarcely more than one fourth the length of the last abdominal somite; the tail-setæ (fig. 8) are three, two very short and one of considerable length, the latter being sharply geniculated above the middle, and equalling in length the last three abdominal segments. Length $\frac{1}{18}$th of an inch (1·4 mm.).

One specimen only of this remarkable Copepod has come under my notice. It was dredged off Hartlepool, in a depth of five fathoms, amongst muddy sand. Without more knowledge of the mouth-apparatus, it is impossible to assign to the species more than a provisional place, but it seems to me not unlikely that it may be found to be of parasitic or semi-parasitic habits. It is very similar, in general character, to the genus *Ophthalmopachus*, Hesse.]

Family 7. CORYCÆIDÆ, *Thorell.*

Body composed of eleven or twelve segments, elongated or subpyriform; abdomen elongated, much narrower than the cephalothorax; head usually anchylosed with the first thoracic segment. Anterior antennæ alike in both sexes, 5—7-jointed; posterior simple, 3- or 4-jointed, forming a prehensile hand, which is armed at the apex with a claw. Mouthorgans (except the second pair of foot-jaws) minute, and destitute, or nearly so, of palps. Posterior footjaw forming a prehensile organ, and, in the male, powerfully clawed. First four pairs of feet alike or nearly so, and adapted for swimming, 2-branched. Fifth pair of feet rudimentary, alike in both sexes, seldom altogether absent. Heart wanting. In addition to the small median eye, there are usually two large lateral eyes, each composed of a single lens. The sexual organs, in both male and female, are double and symmetrical. Ovisacs usually two.

The British genera belonging to this family are *Corycæus*, and *Monstrilla*.

Genus 1. CORYCÆUS, *Dana* (1845).

(Dana, Proc. Acad. Nat. Sc. Philadelph.)

Body elongated, subcylindrical; abdomen 2-jointed, penultimate segment of the cephalothorax produced ventrally into two hook-like processes (Pl. LXXXIV,

fig. 14 a, and Pl. LXXXIII, fig. 11 a a). Anterior
antennæ 6-jointed, short; posterior uncinate, powerfully prehensile, terminal claw much longer in the
male than in the female. Mandibles (Pl. LXXXIV,
fig. 10) slender, divided into two apical teeth (a), and
bearing a minute setiferous palp (b); maxilla (fig.
10 c) composed of a single lamina, which bears several
marginal laciniæ, and near the middle a crescentic
row of small setæ (see also Pl. LXXXIV, fig. 9). First
pair of foot-jaws short and stout, alike in both sexes,
apex produced into a stout hook-like spine, inner
margin bearing several setæ (fig. 11). Posterior
foot-jaws elongated, 3-jointed, forming a strongly
uncinate prehensile hand (Pl. LXXXI, figs, 16, 17),
the claw very much larger in the male. First, second,
and third pairs of feet having both branches 3-jointed,
(Pl. LXXXIV, fig. 12); inner branch of the fourth pair
(fig. 13) rudimentary, 1-jointed; fifth pair (Pl. LXXXI,
fig. 18), rudimentary, composed of a single small
setiferous joint. Last joint of the cephalothorax
(Pl. LXXXIV, fig. 14 b) very small and partially over-
lapped by the penultimate joint. Frontal eyes two,
each composed of a single large, colourless, highly
refracting lens, situated near the bases of the anterior
antennæ. " Conspicilla (lentes frontales) fere unita,
maxima; oculus impar parvulus; oculi superiores
remoti, corpori pigmentato styliformi, plus minusve
curvato " (Claus).

Dana notes the absence of the basal appendages of
the abdomen (fifth pair of feet) as a character of the
genus; these, however, though present in such species

as have come under my notice, are very minute, and have probably on that account been overlooked. By Claus, the inner branches of the swimming-feet are stated to consist of two joints only, but in our British species three joints are distinctly present. The visual organs I have had no opportunity of examining in fresh specimens and must, therefore, accept as correct Dr. Claus' description which, briefly, is as follows :—The frontal lenses are large and very convex, and approach each other towards the median line ; much behind these, in the lower portion of the cephalothorax, lie the elongated rod-shaped pigment bodies, slightly curved and converging towards their bases ; the median eye-spot is single, as in *Sapphirina*, and in the case of *C. germanus* (*anglicus*) shows a pigment-spot in the form of a half *x*, and one (or two ?) crystalline lenses.

1. CORYCÆUS ANGLICUS, *Lubbock*. Pl. LXXXI, figs. 16—19, Pl. LXXXIII, figs. 11—15, and Pl. LXXXIV, figs. 10—14.

> *Corycæus anglicus*, Lubbock. On Eight New Species of Ento-
> mostraca found at Weymouth (Ann. and
> Mag. Nat. Hist., 2nd ser., vol. xx, pl. xi,
> figs. 14—17, 1857). On some Oceanic
> Entomostraca collected by Capt. Toynbee
> (Trans. Linn. Soc., vol. xxiii), p. 182,
> pl. xxix, figs. 10, 11 (1860).
> — *germanus*, Leuckart. Carcinologisches, Archiv für
> Naturg., t. vi, fig. 9 (1859).
> — — Thorell. Bidrag till Kännedomen om Krus-
> taceer, t. xi, xii, fig. 17 (1859).
> — — Claus. Die frei-lebenden Copepoden, p. 156,
> t. ix, figs. 1—4; t. xxiv, figs. 5, 6 ;t. xxviii,
> figs. 1—4 (1863).

Anterior antennæ (Pl. LXXXIII, fig. 12) 6-jointed, robust, about one third as long as the first cephalo-thoracic segment; fourth joint somewhat longer than any of the rest; setæ of moderate length. Posterior antennæ 3-jointed (Pl. LXXXIII, figs. 13, 14), the basal joint bearing two very strong spine-like setæ (a, a); the second joint in the *male* is elongated, sub-quadrangular, has a hooked process (fig. 14 *b*) at the distal angle of the inner margin, and a row of minute serratures along the middle of the joint) (*c*); in the *female* (fig. 13) the joint is ovate and destitute of spines and teeth; the last joint is short, slender, and armed with two pairs of short lateral spines and a large terminal uncinate claw, which in the *male* reaches beyond the extremity of the basal joint. The inner branches of the first three pairs of swimming-feet (Pl. LXXXIV, fig. 12) are 3-jointed, and only about half as long as the outer branches; the outer branches are armed with strong lancet-shaped spines on the outer margin, the last joint very long, bearing two marginal and two apical spines, the latter being long and marginally ciliated (Pl. LXXXI, fig. 19). The basal portion (peduncle) of the fourth pair (Pl. LXXXIV, fig. 13) is elongated and angularly bent, and the inner branch consists of one small bisetose joint; the fifth foot is rudimentary (Pl. LXXXI, fig. 18), consisting of a single minute joint, to which are attached two long slender setæ. The penultimate thoracic segment (Pl. LXXXIV, fig, 14) is produced into a long hook-like angle (*a*); the first abdominal segment has a small anterior hook at its base (*c*), and in the male is elongated and indistinctly biarticulate, but in the female is undivided. The abdomen of the *male*, including the tail-setæ, is equal in length to the cephalothorax; the caudal segments (Pl. LXXXIII, fig. 15) are not quite double the length of the last abdominal segment, the length of both together being just equal to that of the first elongated division of the abdomen; in the *female*, however, the caudal laminæ are longer. The tail-setæ of the *male* are short and stout, the innermost and longest being about half as long as the abdomen, the other two successively shorter; the caudal segments have also on the outer edge, a little removed from the apex, a short slender seta; in the *female* all the tail appendages are shorter and more slender. Length $\frac{1}{28}$th of an inch ('9 mm.). The colour of the animal is a clear glaucous green.

Sir John Lubbock first described this species from specimens taken at Weymouth; Dr. Claus also describes it in his work on the 'Copepoda of the Mediterranean and North Sea,' but without giving the precise locality. Captain Toynbee's specimens were taken in two widely separate localities, viz. lat. 7° 15' N., long. 27° 52' W., and lat. 13° 43' S., long. 33° 55' W. For the gatherings in which the specimens here described were found I am indebted to my friend Mr. E. C. Davison, but although the species occurred in several places—all of them off the Atlantic shores—it was in no case met with abundantly, and although I have myself frequently used the surface-net in those districts I never to my knowledge took a single example of *Corycæus*. The following are the localities of Mr.

Davison's captures :—Between Cornwall and Cape Clear; near Valentia, and off the Skelligs; Dingle Bay, Kinsale and Valentia Harbours. In most, if not all, cases the gatherings were made by the tow-net.

Genus 2. MONSTRILLA, *Dana* (1848).

Body elongated, compressed. Posterior antennæ, and all the masticatory mouth organs, wanting. Branches of all the swimming-feet 3-jointed; fifth foot consisting of one biarticulate branch. Abdomen of the female composed of three segments. Eye large, and situated on the forehead.

1. MONSTRILLA ANGLICA, *Lubbock*.

> *Monstrilla anglica*, Lubbock. On Eight New Species of Entomostraca taken at Weymouth (Ann. and Mag. Nat. Hist., vol. xx, 2nd series, pl. x, figs. 7, 8, 1857).

Of this Copepod I know nothing except from the description referred to above, the specimen itself being unfortunately lost or mislaid. The following is Sir John Lubbock's description of the animal :— " Frons quadrata, angulis rotundatis. Cephalothoracis segmentum primum postice paullo latius. Antennæ 5-articulatæ, setis antenna brevioribus. Abdomen 4-articulatum, segmentis subæqualibus. Styli caudales oblongi, divaricati, setis 6-subæquis, diffusis.

"This species differs considerably from *M. viridis*. In the first place the cephalothorax is rather broadest behind instead of in the middle, and the three posterior segments are somewhat moniliform, so that their sides do not form an even line. The abdomen is 4-jointed, and the basal segment bears on each side a large plumose hair, which passes backward and outward. Upon the fourth caudal seta (counting from the outside) is another, rather smaller than the other five. It is so close to the fourth seta that it might easily be overlooked. For the sake of clearness, however, I have in my figure separated them. Length of·cephalothorax ·037 of an inch, of abdomen ·012, total ·049; of antenna ·027. Caught at Weymouth, October, 1857."

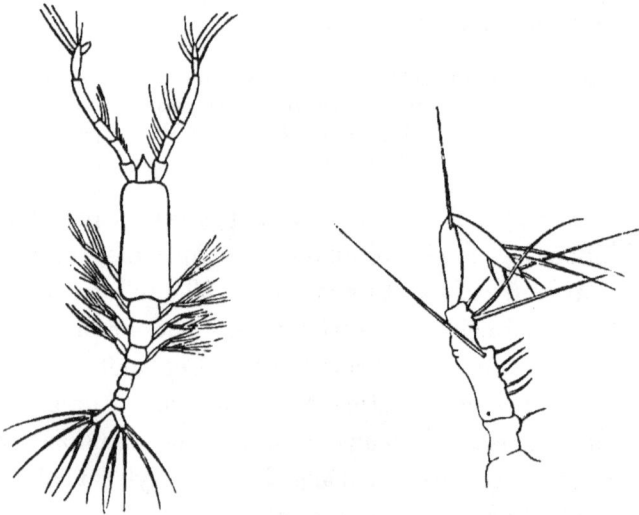

The woodcut here given is copied from Sir John Lubbock's drawings, and represents, in one figure, the entire animal, in the other a more enlarged view of the anterior antenna.

Family 8. SAPPHIRINIDÆ, *Thorell*.

Body composed of eleven or twelve segments, either long or subovate, abdomen broad or subpyriform, cephalothorax subovate; much broader than the abdomen; head usually coalescent with the first thoracic segment. Anterior antennæ composed of five to seven joints; posterior simple, armed at the apex with a claw or several curved setæ. Mandibles small, subulate, with a tapering extremity, or divided into a few slender teeth; maxilla composed of a single small setiferous digit attached near the base of the mandible; first pair of foot-jaws bearing a few spine-like setæ at the apex; second pair bearing a terminal claw, which is much larger in the male than in the female. First four pairs of feet 2-branched, each branch 3-jointed, except sometimes in the case of the inner branch of the fourth foot, which may be only 1- or 2-jointed; fifth pair small, usually 1-jointed, ovisacs two. Animals either free or living in the body cavities of various Tunicata.

The only British genus is *Lichomolgus*.

Genus 1. LICHOMOLGUS, *Thorell* (1859).

(*Boeckia*, Brady, *Macrocheiron*, Brady.)

Body elongated, subpyriform, composed, in the male, of twelve, in the female, of eleven, segments; cephalo-thorax ovate; the head large and coalescent with the first thoracic segment; abdomen slender, first and second segments united in the female, separate in the male. Anterior antennæ 6- or 7-jointed, alike in both sexes, posterior shorter, simple, 3- or 4-jointed, and bearing at the apex a few curved setæ. Mandible forming a slender ciliated stilet, dilated at the base, but excessively slender and filiform beyond the middle; no palp. Maxilla springing from near the base of the mandible, composed of a digitiform process, which bears a few small setæ. First pair of foot-jaws 2-jointed, basal joint large, apical joint slender and armed with a few terminal and marginal setæ, some of which are spine-like; second pair 2-jointed, robust, in the female bearing three or four spine-like setæ, in the male a large uncinate apical claw. First four pairs of feet having both branches 3-jointed, except the fourth, which has the inner branch 1- or 2-jointed; fifth pair of feet small, 1-jointed, rudimentary.

Animals living either free or in the branchial sacs of various simple Ascidians.

1. LICHOMOLGUS FUCICOLUS, *Brady.* Pl. LXXXV, figs.
1—11.

Macrocheiron fucicolum, Brady. Nat. Hist. Trans. Northumberland and Durham, vol. iv, p. 434,
pl. xviii, figs. 9—18 (1872).
Lichomolgus fucicolus, Brady & Robertson. Ann. and Mag. Nat.
Hist., ser. 4, vol. xii, p. 140 (1873).

Body elongated, first segment equal to more than
half the length of the cephalothorax, which tapers
gradually to its last joint; the abdomen in both sexes
is composed of five joints (figs. 1, 10), but in the *male*
the first joint has its lower angles produced into two
short sharp spines, and is much enlarged both in
length and breadth, being as long as the rest of
the abdomen, including the caudal segments; in
the *female* the first abdominal segment is formed
by the union, often imperfect, of the first and second
somites (the line of separation between the two being
quite visible in immature specimens), and, like that of
the male, is about equal in length to the remaining
half of the abdomen, the angles of the true first somite
at the line of junction being, however, smoothly
rounded off. Rostrum short and sharp; anterior
antenna (fig. 2) slender, 7-jointed, nearly as long as
the first body segment, second, fourth, and fifth joints
longer than the rest, third joint shortest; the whole
limb beset at the apex and on the outer margin with
rather short slender hairs. Posterior antenna (fig. 3)

3-jointed, bearing a few short marginal setæ, and at
the apex of the third joint four long rigid setæ and a
large falciform claw, which is denticulated on its
inner margin, and equal to half the length of the
antenna. The mandible (fig. 4 *a*) is small, pyriform,
dilated at the base, and produced into a long stilet-
shaped extremity, which is ciliated on both margins;
the maxilla (fig. 4 *b*) forms a short digitiform process
attached near the base of the mandible, and bearing
four slender apical setæ. Basal joint of the anterior
foot-jaw (fig. 5) large and stout, bearing a single
curved and ciliated seta (fig. 5 *a*) at the distal extremity; terminal joint (*b*) produced laterally into a
slender apical seta, and pectinated at the base with
fine teeth, which become gradually finer and more hair-
like towards the apex. Posterior foot-jaws 2-jointed,
the last joint in the *female* (fig. 6) short and broad,
bearing a short, robust, curved spine at the apex, and
on the margin two or three more slender spines; in
the *male* (fig. 7) the second joint is elongated, oval,
bearing a fringe of hairs and a single stout curved
spine on the inner margin, and at the apex an extremely long falciform claw, the apex of which reaches
as far as the middle of the basal joint. The joints of
the inner branches of the swimming-feet are, in the first
three pairs, produced into short spines at the external
angles; the marginal spines of the outer branches are,
as in other species of *Lichomolgus*, lancet-shaped, the
central axis being bordered by an almost pellucid
membranous margin; the fourth pair (fig. 9) has the
outer branch much elongated, the spines and setæ

being proportionately long and slender; the inner branch 2-jointed, scarcely exceeding in length the first joint of the outer branch, and bearing only two setæ at the apex of the second joint. Fifth pair of feet (fig. 10 a), alike in both sexes, composed of a single long curved joint with two apical setæ. The caudal segments are parallel, subcylindrical, about thrice as long as broad, and equal in length to the last two abdominal segments; the setæ are four on each segment, two short, lateral or subapical, and two apical, which are much longer, the innermost of the two being equal to about two thirds the length of the abdomen; these longer setæ are jointed and suddenly constricted in the middle. The colour of the animal is usually a deep brown, but varies with habitat. Length of the male $\frac{1}{25}$th of an inch (1 mm.), of the female $\frac{1}{20}$th of an inch (1·3 mm.).

Lichomolgus fucicolus inhabits chiefly the littoral and Laminarian zones, but occurs not unfrequently in greater depths; it is a very well marked and, in the British seas at least, a widely distributed species, as will be evident from the following list of localities in which I have taken it:—Amongst fuci, near low water-mark, on the coasts of Northumberland and Durham (St. Mary's Island and Ryhope), at the end of summer; dredged four miles off Hawthorn and Marsden (Durham), amongst rough shelly sand, in a depth of about twenty-five fathoms; off the Yorkshire coast near Scarborough, Robin Hood's Bay, and Red Cliff, in depths of about thirty-five fathoms; amongst weeds at Clifden, Roundstone, and Westport (Ireland); and in

Lough Swilly, amongst muddy sand, from a depth of eight fathoms.

This species is interesting as being an example of an entirely free-living *Lichomolgus*, about whose usual haunts there can be no doubt. I have not seen a single specimen in any collection of ascidicolous species, and from its usual dark brown colour and its frequent occurrence on the fronds of fuci I think that its nourishment is, in all probability derived either from the juices of algæ, or from still more minute animals living upon their fronds.

The differences between this and the typical forms of *Lichomolgus* do not seem important enough to warrant the retention of the generic term *Macrocheiron*, which I at one time applied to it. The 7-jointed anterior antenna is met with in some other species (*L. liber*), while the long-clawed, 3-jointed, posterior antenna may be accepted as of not more than specific importance.

2. LICHOMOLGUS LIBER, *Brady* and *Robertson.* Pl. LXXXVI, figs. 1—13.

Lichomolgus liber, B. & R. Brit. Assoc. Report, p. 197 (1875).

In general appearance like the last species, but somewhat more robust; joints of the first pair of antennæ (fig. 2) stouter, the last joint much the shortest, and only about one fourth as long as the second; third and sixth joints about once and a half

as long as the last. The basal portion of the mandibular stilet (fig. 4) bears a comb of fine tooth-like hairs
on its outer margin; the marginal pectination of the
first foot-jaw (fig. 5) is of a finely setose character; the
second foot-jaw of the *female* (fig. 6) has a subpyriform
terminal spine; that of the *male* (fig. 7) has an elongated subovate hand, which is fringed with short cilia
along the proximal half of its inner edge, and is armed
at the apex with a long falciform claw. The swimming-
feet are not very different from those of the preceding
species, except the fourth (fig. 10), which has its inner
branch uni-articulate, and as long as the first two joints
of the outer branch. The fifth foot (fig. 11) consists
of a single slender, subulate joint, to the base of which
is attached a stout seta, equal in length to the joint
itself. The abdomen of the *female* (fig. 12) consists
of five segments, the first and second being distinctly
separate; in the *male* (fig. 13) the first segment is
much enlarged, and produced backwards into two
strong lateral spines; all the joints in both sexes are
much broader than long; the caudal segments are
about twice as long as broad, and equal in length to the
last abdominal somite; terminal setæ five, subequal,
finely plumose, and about half as long as the abdomen.
Length $\frac{1}{22}$nd of an inch (1·1 mm.).

I have notes of the occurrence of this species in
dredgings from Marsden and Hawthorn (Durham
coast), 20—27 fathoms; off Scarborough; off the
south end of the island of Bute, 16 fathoms; and
amongst muddy sand, Lough Swilly, in a depth of
7—8 fathoms. As in the case of *L. fucicolus,*

I do not know of its occurrence in the cavities of Ascidians.

The distinctly divided first abdominal segment of the females of this species, though unmistakeably observed in all which I have dissected, may very probably be a sign of immaturity, none of my specimens having been found with ova; and I suspect also that the undivided inner branch of the fourth foot may be explicable in the same way.

3. LICHOMOLGUS ARENICOLUS, *Brady.* Pl. LXXXVII, figs. 1—7.

Boeckia arenicola, Brady. Nat. Hist. Trans. Northumberland and Durham, vol. iv, p 430 (1872).

Body elongated, subpyriform; anterior antennæ about half as long as the first segment of the body, 6-jointed (fig. 2), second joint much the longest, third less than one third as long, fourth, fifth, and sixth nearly equal, and about two thirds the length of the second joint. Mandibles (fig. 3) much like those of *L. liber.* Last joint of the first pair of foot-jaws (fig. 4) armed with six nearly equal marginal spines. Last joint of the second foot-jaw forming, in the *male* (fig. 5), a broadly ovate hand, roughened with numerous denticulations, and bearing a long terminal falcate claw. The fifth foot (fig. 7) is in the form of a long subclavate joint, bearing on the broad truncate apex three setæ, one of which is short, the other two nearly as long as the limb itself;

the outer margin has a long spine at a little distance from the apex, and is fringed with closely-set cilia; the inner margin has, near the middle, a series of five or six stout setæ. The abdomen is slender, the first segment swollen, and twice as long as any of the rest, second, third, and fourth segments about as long as broad; caudal laminæ longer than the last abdominal segment, and about thrice as long as broad, bearing one short seta about the middle of the outer margin, and three at the apex, of which two are long and one very short, the innermost about equal in length to the four abdominal segments. The ovisacs of the female are large and divergent, reaching nearly to the extremity of the abdomen. Length $\frac{1}{13}$th of an inch (1·9 mm.).

I have seen only three specimens of this species, one of which was dredged off Battery Point, Cumbrae, one off Robin Hood's Bay, in a depth of thirty fathoms, and one off Seaton Carew (Durham), in four fathoms, on a sandy bottom. There can be no doubt of its specific distinctness, but various details of structure remain yet to be worked out.

4. LICHOMOLGUS THORELLII, *Brady* and *Robertson*. Pl. LXXXVIII, figs. 1—9.

Lichomolgus Thorellii, B. & R. Brit. Assoc. Report, p. 197 (1875).

Anterior antennæ (fig. 1) 7-jointed, the last joint very small, third about twice as long as the seventh.

The maxilla (fig. 3) is bisetose ; the anterior foot-jaw
(fig. 4) has a broad basal and an attenuated apical
joint, which bears four stout terminal setæ, and at the
base one long ciliated spine and two shorter spines.
The last joint of the posterior foot-jaw is in both sexes
short and broadly ovate, in the *female* bearing four
stout spines (fig. 5), in the *male* (fig. 6) a fringe of marginal setæ and a long curved terminal claw.   The
inner branch of the fourth foot (fig. 7) is 1-jointed,
stout, nearly as long as the entire outer branch, and
has a truncate extremity, from which spring two long
and stout spines, and which is produced into a sharp
tooth at the internal angles; the margins densely and
finely ciliated.   Fifth pair of feet uncinate, bisetose.
The abdomen of the *female* (fig. 8) is very slender, the
first joint considerably longer than the following three,
third joint very small, only about half as long as broad,
second joint twice as long as the third, fourth nearly
twice as long as the second.   The tail-segments are
excessively long and narrow, equal in length to the
first abdominal somite, and about ten times as long as
broad; the external margin bears one small seta near
its middle, and the apex four or five slender setæ, one
of which is nearly as long as the entire abdomen
(tail-segments included), another about half as long,
the rest being quite small.   The first abdominal
segment of the *male* (fig. 9) is as broad as long, truncate behind, its posterior angles each bearing a stout
spine and two or three short setæ ; the remaining four
segments are from side to side only about one third
the bulk of the first, otherwise they and the caudal

segments differ scarcely at all from those of the
female.

I have not met with this species except off the coasts
of Durham and Yorkshire, in which district I have
memoranda of its occurrence as follows :—Off Marsden,
twenty-five fathoms, amongst coarse, shelly sand, one
specimen only; off Hawthorn, twenty-seven fathoms,
one specimen ; a few in a dredging, chiefly of *Filagrana
implexa*, in thirty-five fathoms, off Robin Hood's Bay.

5. Lichomolgus furcillatus, *Thorell*.   Pl. LXXXVIII,
figs. 10—14.

Lichomolgus furcillatus, Thorell.   Om Krustaceer i ascidier, p. 74,
tab. xiii, fig. 20 (1859).

The anterior antennæ (fig. 10) are 6-jointed, the
second joint once and a half as long as the third ;
third, fourth, and fifth equal ; last joint less than one
half the size of the preceding.   Mandibles (fig. 11)
very slender, and but slightly ciliated near the base.
Inner branch of the fourth foot 2-jointed.   Fifth foot
(fig. 14 *a*), narrow, oblong, bearing two apical setæ.
First abdominal segment in the *female* nearly as long
as the three following segments ; second, third, and
fourth equal, not quite as long as broad ; caudal appendages twice as long as the last segment, and about
four times as long as broad, bearing one short seta
beyond the middle of the outer margin, also four apical
setæ, two of which are short, and two long, the

longest being in length about equal to the abdomen. Length $\frac{1}{16}$th of an inch (1·5 mm.).

A few specimens only of this species have been found, and though they differ in some respects from those of *L. furcillatus,* as figured by Thorell, more especially in the characters of the caudal segments and the fifth pair of feet, the differences do not seem sufficiently pronounced to warrant their separation from that species. The British habitats are the following :—In the branchial sac of *Corella parallelogramma,* Shetland (from specimens collected by the Rev. A. M. Norman) ; taken in the tow-net in Roundstone Bay and Westport Bay ; dredged in a depth of 7—8 fathoms, amongst muddy sand, in Lough Swilly ; and in Mulroy Lough (Donegal), ten fathoms. Single specimens only were found in each case.

6. LICHOMOLGUS FORFICULA, *Thorell.* Pl. LXXXV, figs. 12—16, and Pl. LXXXVI, figs. 14—18.

Lichomolgus forficula, Thorell. Om Krustaceer i ascidier, p. 73, tabs. 12, 13, fig. 19.

Anterior antennæ (Pl. LXXXVI, fig. 14) 6-jointed, first three joints of equal length, about thrice as long as broad, fourth rather shorter, fifth half as long as the third, sixth half as long as the fifth ; second and third joints clothed with unusually long marginal hairs. Posterior antennæ 3-jointed, having two stout

uncinate terminal claws in place of the usual brush of
setæ. Mandible (Pl. LXXXV, fig. 12) densely clothed
at the base with long marginal setæ, those of the outer
side the longest, and densely plumose ; maxilla digitiform, bisetose. Spines of the anterior foot-jaws long
and slender (Pl. LXXXVI, fig. 16), each bearing a
series of long setæ on the external margin near the
base ; last joint of the second foot-jaw, in the *female*,
short and broad (fig.17), bearing a single short uncinate
claw ; in the *male* (Pl. LXXXV, fig. 13) subovate,
and ending in one short and one very long falcate
claw. Inner branch of the fourth swimming-foot
(fig. 14) 2-jointed, first joint short and bearing a
single seta, second nearly twice as long, somewhat
pear-shaped, with a broad truncate apex, from which
spring two spine-like setæ ; fifth foot consisting of a
single joint with two terminal setæ. First abdominal
somite of the *male* (fig. 16) much swollen, rounded
below, the other segments all very short ; caudal
appendages slender, jointed in the middle, as long as
the last five abdominal segments, terminal setæ rudimentary ; in the *female* (Pl. LXXXVI, fig. 18) the
second and fifth abdominal somites are much elongated,
and the first is scarcely at all tumid.

In the " Last Report of Dredging among the Shetland Isles," Mr. Norman records this species as having
been found in the " water passages and branchial sac
of *Ascidia mentula*." It occurs also in a collection of
Copepoda found by Mr. Norman in Ascidians dredged
at Oban and kindly sent to me by him for examination ; and I have myself found it amongst specimens

dredged in a depth of ten fathoms, on a bottom chiefly of nullipores, in Mulroy Lough, Donegal.

Section III.—SIPHONOSTOMA, *Thorell.*

Fam. 9. ARTOTROGIDÆ, *nov. fam.*

Body broad, depressed, rounded or subovate, composed of 10—12 segments, first segment very large, and composed of the coalescent cephalic and first thoracic somites. Anterior antennæ 9—20-jointed, alike, or nearly alike, in both sexes; posterior short, composed of three or four joints, sometimes bearing a small 1-jointed secondary branch, and at the apex a claw or a few curved setæ. Mouth produced into a siphon composed of the elongated labrum and labium; mandibles stilet-shaped, simple or provided with a filiform palp; maxillæ usually 2-branched and setiferous; first and second pairs of foot-jaws simple, prehensile, clawed at the apex, 2—4-jointed. First four pairs of feet 2-branched (except the fourth pair of *Dyspontius*), each branch 3-jointed. Fifth pair small and composed of one or two joints, or altogether wanting.

Animals usually free, but sometimes living in the cavities, or on the integument, of various marine Invertebrata.

The British genera are *Cyclopicera, Acontiophorus, Artotrogus,* and *Dyspontius.*

Genus 1. CYCLOPICERA, *Brady* (1872).

Cephalothorax broadly ovate; abdomen and last segment of thorax much narrower; head united with first segment of thorax; abdomen of the male 4-, of the female 3-jointed. Anterior antenna shorter than the first somite, 19- or 20-jointed; posterior 3-jointed, and provided with a very small secondary branch. Labrum and labium produced into a short and wide suctorial mouth. Mandible stilet-shaped, and provided with a simple filiform palp; maxilla two branched, the branches slender, digitiform, and setiferous at the apex. Anterior foot-jaws 2-, posterior 4-jointed, nearly alike in general shape and in size, strongly clawed and adapted for prehension. First four pairs of feet short and stout, both branches composed of three joints; fifth pair small, 2-jointed; ovisacs two.

The many-jointed antennæ, together with the very well marked characters of the mandibles, maxillæ, and foot-jaws, separate this genus at once, and very decidedly, from the foregoing. It seems, indeed, to occupy a position intermediate between *Lichomolgus* and *Acontiophorus*, approaching the latter genus in the characters of the antennæ and swimming-feet, but differing considerably as regards the mandibles and the extent of development of the siphonal mouth. From *Artotrogus* it differs only in the character of the mandibles (the extremities of which show an approach

to the ordinary toothed character, and are also provided with a palp), and in the less elongated suctorial
mouth. It may be doubted, however, whether further
examination may not show the propriety of uniting
both genera under *Artotrogus* (*Ascomyzon*, Lilljeborg).

1. CYCLOPICERA NIGRIPES, *Brady* and *Robertson*. Pl. LXXXIX, figs. 1—11.

Cyclopicera nigripes, B. & R. Brit. Assoc. Report, p. 197 (1875).

Robust; cephalothorax broadly ovate (fig. 1); abdomen short and stout, the first and second somites
having their posterior margins produced into sharp
lateral angles. Anterior antennæ (fig. 2) composed of
nineteen joints, the first the largest, the second to the
ninth at least twice as broad as long, tenth to sixteenth narrower, and about as broad as long, seventeenth and nineteenth fully twice as long as broad,
eighteenth about one third as long as the nineteenth;
the first joint has at its apex a long sharp spine, the
other joints each bearing one or more very short setæ,
besides which the ante-penultimate joint gives attachment to a sword-shaped olfactory appendage, and the
last bears four or five long apical setæ. First joint of
the posterior antennæ (fig. 3) long, and bearing a
minute 1-jointed secondary branch; second joint
about half as long as the first; third still shorter,
bearing two small setæ and a long, slender, curved
claw. Mandible (fig. 4) stout, stilet-shaped, almost

imperceptibly serrated at the apex; palp slender, bearing one very long and one short terminal seta. Maxilla (fig. 5) 2-branched, each branch elongated, digitiform, simple, and armed apically with three long and stout setæ. Anterior foot-jaw (fig. 7) 2-jointed, and armed with a long doubly-curved claw; second pair (fig. 8) longer, 3-jointed, with a long and slender falciform claw. First four pairs of feet (fig. 9) short and stout, nearly equal in size, both branches 3-jointed; fifth pair (fig. 11 *a, a*) 2-jointed, the terminal joint broadly ovate, bearing two marginal spines and three apical setæ. First abdominal somite in both sexes large, quadrangular, sharply emarginate in the middle, and having its lateral angles produced into sharp spines posteriorly; the second somite is about half as long and similarly angulated, and so also is the third in the male, though not in the female; the second and third somites in the male are equal in length, but in the female the second is considerably shorter; the fourth somite in the male is short, being about equal in length to the caudal segments, which are stout, and a little longer than broad; terminal setæ four in the male, five in the female, finely plumose, two about as long as the abdomen, the rest short. Colour pale brown, the feet smoky black, a character which at once distinguishes the animal amongst others. Ovisacs two, large, nearly spherical, and divergent. Length $\frac{1}{20}$th of an inch (1·2 mm.).

This fine and very distinct species is generally distributed, but seems to occur in greatest abundance on the north-east coast of England. I have notes of its occurrence in the following dredgings:—Off Marsden and Hawthorn (Durham), twenty-five to twenty-seven fathoms, on rough sand; three miles off Robin Hood's Bay (Yorkshire), on rough sand, thirty fathoms; five miles off the same place, amongst shells and gravel, thirty fathoms; and under similar conditions off Red Cliff, Staiths, and Scarborough, in the same district; in washings of *Laminariæ* from Hillswick, Shetland (*Rev. A. M. Norman*); dredged in the Firth of Clyde off Callum's Hole (Bute); off Portincross (Ayrshire), fifteen fathoms, and in Lough Swilly, seven fathoms; one or two specimens taken also off Sunderland in the surface-net.

2. CYCLOPICERA LATA, *Brady.* Pl. LXXXIX, fig. 12; and Pl. XC, figs. 11—14

Cyclopicera lata, Brady. Nat. Hist. Trans. Northumberland and Durham, vol. iv, p. 433, pl. xviii, figs. 3—8 (1872).
Ascomyzon echinicola, Norman. Last Report of Dredging among the Shetland Isles (Brit. Assoc. Report, p. 300, 1868).

In the *female* the anterior antennæ (Pl. XC, fig. 11) are 20-jointed, the last joint shorter than the penultimate; in the *male* (fig. 12) 17-jointed, the eleventh, twelfth, and thirteenth, and the sixteenth and seventeenth joints being confluent; the fifth foot (Pl. LXXXIX, fig. 12) is small, 1-jointed, and has two apical setæ; the first joint of the female abdomen is rounded at the angles and fringed with a few short cilia on the middle of each margin ; caudal appendages about as long as the last abdominal segment; longest setæ equal in length to the abdomen.

One specimen was found amongst algæ in tidal pools at Roker, near Sunderland; and a few were dredged in Lough Swilly, in a depth of about eight fathoms, amongst muddy sand. Found also by the Rev. A. M. Norman upon *Echinus esculentus* in Shetland.

This does not differ very materially from *C. nigripes*, except in the number of antennal joints and the character of the fifth pair of feet and abdominal segments, which are rounded and not at all angular. The animal is also less robust and colourless. I am not at all sure that it may not be an immature form of *Artotrogus (Ascomyzon) Lilljeborgii*, the development of which has not, as yet, been accurately observed.

Ascomyzon echinicola, Norman, while differing in form of mouth and mandibles from the typical *Ascomyzon*, seems to agree with *Cyclopicera lata*, so far as I can make out from the type-specimens kindly lent me by Mr. Norman. In this case, though the specific name *echinicola* has the claim of priority, it is, I think, inadmissible, the animal being certainly by no means always, or even, perhaps, very commonly, parasitic on Echini. For the present, at any rate, it seems to me best to consider the species identical with *C. lata*, and to accept that name.

3. CYCLOPICERA GRACILICAUDA, *nov. sp.* Pl. LXXXIII, figs. 1—10.

Anterior antennæ (fig. 1), slender, 20-jointed, first ten joints much shorter than broad, eighteenth and twentieth joints the longest, being three or four times as long as broad; the whole limb is sparingly clothed with slender setæ on its outer margin; the posterior antenna (fig. 2) has one slender terminal claw, and a minute secondary branch attached near the middle of its first joint; the extremity of the mandible is distinctly divided into teeth (fig. 3); fifth foot (fig. 10) composed of two joints, the first of which is small and has a small spine-like process at its distal angle, the second curved, elongated, about five times as long as broad, and bluntly serrated at the apex. Abdomen slender, the first joint elongated; caudal segments very long and slender, length nearly equalling that of the first two abdominal somites, and ten or twelve times greater than the breadth; the principal seta is about as long as the abdomen.

A single specimen of this species was dredged in a depth of thirty-five fathoms off Robin Hood's Bay, Yorkshire.

Genus 2. ARTOTROGUS, *Boeck* (1859).

(*Artotrogus, Asterocheres,* Boeck, Tvende nye parasitiske Krebsdyr, 1859.
Ascomyzon, Thorell, Om Krustaceer i Ascidier, 1859.)

Body broad, suborbicular or subpyriform; cephalo-thorax broadly ovate, composed of eleven somites, the first of which is united with the head; abdomen composed of four segments, the first and second of which are, in the female, coalescent. Anterior antennæ composed of nine, ten, or twenty joints, shorter than the first body segment; posterior 4-jointed, bearing a strong apical claw, unbranched or furnished with a very small secondary branch. Mouth produced into a siphon, which reaches near or beyond the hinder margin of the first segment of the cephalothorax. Mandible elongated, filiform, without a palp. Maxillæ 2-branched, setiferous at the apex. First and second pairs of foot-jaws simple, bearing a strong apical claw, first composed of two, second of four joints; first four pairs of feet having both branches 3-jointed; fifth pair rudimentary, composed of one joint only. Animals living in the branchial sacs of simple Ascidians or on the integument of various marine Invertebrata.

The three genera, *Ascomyzon,* Thorell, *Asterocheres,* and *Artotrogus,* Boeck, appear to differ in no essential character. The smaller number of antennal joints found in *Artotrogus* cannot be looked upon as a character of generic importance, yet this, together with the more rounded outline, is really the only definable distinction. With this view I have here discarded the name *Ascomyzon*, which, though published in the same year with *Artotrogus*, seems to have been a little later, M. Boeck's pamphlet being referred to in M. Thorell's work on the ' Crustacea inhabiting the Interior of Ascidians.' *Asterocheres* is set aside as being generically synonymous with the less objectionable term *Artotrogus*.

1. ARTOTROGUS BOECKII, *Brady.*   Pl. XCI, figs. 1—9.

Ascomyzon Lilljeborgii, Thorell,   Om Krustaceer i Ascidier, p. 78, tab. xiv, fig. 21 (1859).

Anterior antennæ 20-jointed, exactly similar in structure to those of the preceding genus. Mandibles simple, produced into a long filiform seta, and destitute of a palp (Pl. XCI, fig. 3). Fifth pair of feet ciliated on the margins, and bearing two apical setæ. First abdominal segment fringed below the middle on each margin with a series of about ten short setæ. Caudal segments about as long as broad; setæ five, the larger ones about as long as the abdomen and finely plumose.

I know this species only from two or three specimens taken in the surface-net, and amongst weeds, at a depth of two fathoms, in Westport and Roundstone Bays, Ireland. M. Thorell's specimens were obtained from *Ascidia parallelogramma*, in which he states that it frequently occurs. I have not myself seen any examples taken from Ascidians.

The specific name *Lilljeborgii* having been used by M. Boeck (prior to the publication of Thorell's work) to designate an *Asterocheres*, which genus is, as I think, identical with *Artotrogus*, I propose in this case to substitute the word *Boeckii* for the unavoidably disused *Lilljeborgii*.

Possibly the organs here called mandibles might more properly be looked upon as palps only, in which case the mandible proper is altogether absent. These organs, whatever they are, are placed quite outside of the base of the siphon, as is plainly shown in fig. 6, and, except in *A. magniceps*, I have not found any trace of a mandible included within the siphon itself. Further research may show that the species here asigned to *Artotrogus* include two distinct types of structure as to the mouth organs, but I have not been able to satisfy myself on this point, owing to lack of specimens for dissection.

2. ARTOTROGUS MAGNICEPS, *nov. sp.* Pl. XCIII, figs. 1—9.

Artotrogus orbicularis, Brady and Robertson. Brit. Assoc. Report, p. 197 (1875). (Not *Artotrogus orbicularis*, Boeck.)

First segment (fig. 1) equal in length to more than half of the whole body, width equal to two thirds the length of the body; second, third, and fourth segments narrow, and produced laterally into acutely angular wing-like processes, which are directed backwards; all

the segments separated from each other by deep inden-
tations; the joints of the abdomen, except the first, are
extremely small, the whole abdomen being only equal
in length to the second and third thoracic segments;
caudal segments scarcely longer than broad; apical
setæ five, three very short, the longer of the two
principal equal to about half the length of the body.
Anterior antenna 10-jointed, slender, of about equal
width throughout (fig. 2), almost destitute of setæ
except at the apex, which is furnished with several
small, and two or three moderately long, setæ, and
also with a slender olfactory appendage; the relative
lengths of the joints as in the following formula :

1,	2,	3,	4,	5,	6,	7,	8,	9,	10.
10	17	3	3	8	3	9	6	7	15

The extremities of the two limbs of the siphon are
somewhat lancet-shaped (fig. 3 *a*, *a*) and ciliated, and
the end of the mandible (*b'*) is denticulated on one side.
The anterior foot-jaw (fig. 5) has a small subsidiary
claw near the apex, and the second joint of the posterior pair (fig. 6) is ciliated along the outer margin.
Maxilla composed of two very slender digits (fig. 4),
each of which has two apical setæ, the shorter of the
two being marginally ciliated. The outer margins of
the first and second joints of the swimming-feet are
straight and fringed with closely-set short cilia, the
angles armed with two stout lancet-shaped spines, the
joints constricted at the proximal, and dilated at
the distal, extremity; last joint ovate, bearing three
spines on the outer, five long setæ on the inner margin,
and a long finely denticulated spine at the apex; the

inner branch of the fourth foot is small, almost destitute of setæ, and terminates in two very small spines
(fig. 7). Fifth pair of feet (fig. 8) 1-jointed, quadrate,
bearing one long and two short apical setæ. Shell
structure (fig. 9) areolated, finely punctate.

One specimen dredged in a depth of fifteen fathoms
off Portincross, Ayrshire, and one in thirty fathoms off
Red Cliff, Yorkshire.

3. ARTOTROGUS NORMANI (*Brady* and *Robertson*). Pl. XCI,
figs. 12—15; Pl. XCII, fig.
14; and Pl. XCIII, fig. 10.

Dyspontius Normani, B. & R. Brit. Assoc. Report, p. 197 (1875).

This is, in general appearance, much like the preceding species; the abdomen of the *male* (Pl. XCIII,
fig. 10) is elongated, and has the last joint dilated
distally and longer than either the second or third
joints; caudal appendages wide, and shorter than the
last abdominal joint, each bearing four subequal plumose setæ, the longest of which is not more than equal
to the last three abdominal somites. Anterior antenna (Pl. XCI, fig. 12) 9-jointed, stout, its outer
margin and apex bearing numerous short subequal
setæ; the relative lengths of the joints are as follows:

1,	2,	3,	4,	5,	6,	7,	8,	9.
8	4	11	4	4	4	3	5	8

Posterior antenna simple, composed of four nearly
equal joints (fig. 13), and bearing three small terminal

spine-like setæ. Second pair of foot-jaws (fig. 14) powerfully clawed, and armed with four stout curved spines on the inner aspect, one of these springing from near the middle of the second joint. Fifth pair of feet apparently wanting. The shell structure (Pl. XCII, fig. 14) exhibits numerous circular elevations, each of which bears a minute seta on its summit.

This seems to agree with the genus *Dyspontius* in every respect except that the fourth pair of feet are 2-branched. Though it approaches very closely in most of its characters to *A. magniceps*, it must, I think, be considered quite distinct by virtue of its more densely setose, fewer-jointed, and stouter anterior antennæ, the median spine of the second joint of the posterior foot-jaw, and the characters of the caudal segments and setæ; the shell-structure, also, and the joints of the swimming-feet, differ somewhat. A few specimens only were dredged on the Durham coast, six miles off Hawthorn, in a depth of twenty-seven fathoms.

4. ARTOTROGUS LILLJEBORGII, *Boeck*.

Asterochcres Lilljeborgii, Boeck. Tvende nye parasitiske Krebsdyr, p. 6, tab. ii, figs. 1—11 (1859).

The siphon here is much less elongated than in the foregoing species, the anterior antennæ are slender and 18-jointed, and the fifth pair of feet are 2-jointed. The cephalothorax is very wide in proportion to its length.

Two or three specimens, undoubtedly belonging to this species, I obtained from a sponge dredged in Westport Bay, Ireland, but the dissections which I made are unfortunately not in a condition good enough to allow of minute description. Nor would any good end be served by transferring M. Boeck's description to these pages. The siphonostomous Copepoda require and will well repay a laborious investigation, a work of no very great difficulty, if only a liberal supply of specimens could be obtained. Those which are to be found in the simple Ascidians might probably be obtained readily in sufficient numbers; but in the case of the free-living species, seeing that they are but seldom found, and then must be picked out with great labour from a mass of dredged or net-collected Microzoa, the impediments are rather serious. Except *Cyclopicera nigripes*, which occurred in some plenty amongst material dredged off the Yorkshire coast, I have rarely met with more than two or three examples in a single gathering.

Genus 3. DYSPONTIUS, *Thorell*, 1859.

Like *Artotrogus*, except that the posterior antennæ consist of only three joints, the fourth pair of feet are destitute of an internal branch, and the fifth pair are altogether wanting. The abdomen contains, in both sexes, one segment more than in *Artotrogus*.

1. DYSPONTIUS STRIATUS, *Thorell.*   Pl. XCII, figs. 1—13.

Dyspontius striatus, Thorell.   Om Krustaceer i Ascidier, p. 81
tab. xiv, fig. 22 (1859).

Anterior antennæ (fig. 2) 9-jointed in the female (10-jointed, Thorell), 11-jointed in the male (fig. 3); the following formulæ represent the relative lengths of the antennal joints in the two sexes; female :

$$\frac{1, \quad 2, \quad 3, \quad 4, \quad 5, \quad 6, \quad 7, \quad 8, \quad 9;}{5 \quad 7 \quad 1 \quad 2 \quad 1\frac{1}{2} \quad 2 \quad 2 \quad 3 \quad 5}$$

male :

$$\frac{1, \quad 2, \quad 3, \quad 4, \quad 5, \quad 6, \quad 7, \quad 8, \quad 9, \quad 10, \quad 11;}{5 \quad 6 \quad 1 \quad 1 \quad 1\frac{1}{2} \quad 1 \quad 1 \quad 4 \quad 2 \quad 3 \quad 1}$$

all the joints are sparingly setose on the outer margin, the last bearing, in addition to several apical setæ, a long olfactory rod ; in the *male* the eighth joint has a small blunt hook on the outer margin ; the last joint of the *female* antenna is marked near the apex with several surface irregularities, which obscure its structure, but it appears to be undivided. Posterior antennæ (fig. 4) 3-jointed, the first joint bearing a very minute, almost obsolete, secondary branch ; last joint armed with three apical spines, that in the middle long and strong, the outer scarcely half as long, and the innermost extremely small. Maxillæ (fig. 5) composed of two slender digits, one bearing two, the other one long, apical seta ; mandibles and foot-jaws like those of *Artotrogus*. The first three pairs of swimming-feet have both branches 3-jointed, the marginal spines of the outer branches small, lancet-shaped, and bordered with extremely minute cilia; the first joint of the inner branch has a spine and a long seta near the middle of its inner edge, the second a spine and three setæ, as well as four apical spines, two of which are minute ; the third joint also has two spines on its outer margin. The peduncle of each swimming-foot is produced internally into a very large and prominent rounded angle (figs. 8, 9, a, a). Fourth pair of feet (fig. 9) elongated, 1-branched, destitute of marginal spines; first and second joints with one seta, third with four setæ and a long terminal spine. No fifth foot. First abdominal segment bearing at the posterior angles three slender spine-like setæ (fig. 10). Abdominal somites, except the first, much broader than long; caudal segments scarcely longer than broad, about equal to the last ring of the abdomen; setæ five, the longest about equal in length to the abdomen, the next half as long, the rest very small. The shell is semi-transparent, rather thick, corrugated (fig. 13), and more or less sparingly beset with minute prickles on some portions of the cuticular surface. Length $\frac{1}{14}$th of an inch (1·8 mm.).

This, though widely distributed, seems to be a very scarce species; in most of the localities here noted it was represented only by a single example. On weeds at Tobermory and Hillswick (Shetland), (*Rev. A. M. Norman*); dredged in a depth of forty fathoms off St. Agnes, Scilly Islands; in twenty-seven fathoms off Hawthorn (Durham); in thirty-five fathoms off Red Cliff and Robin Hood's Bay (Yorkshire).

The number and proportions of the antennal joints

in the specimens here described, differ somewhat from
those given by Thorell, who had seen only one specimen, a female. The accuracy of my own drawings I
have verified by reference to several specimens, and it
is, perhaps, better for the present simply to amend
M. Thorell's description than to introduce another
specific name on the strength of these not very important differences. M. Thorell asserts dubiously the 1-
branched character of the fourth foot; of this character
there can be no doubt, nor, so far as I have been able to
ascertain, of the entire absence of the fifth pair. The
Scandinavian specimen on which the genus is founded,
though taken in the open sea by Prof. Lilljeborg, is
nevertheless supposed by M. Thorell to be a true
internal parasite. In this supposition, however, I
cannot concur. Though truly suctorial in structure
these creatures may very likely obtain supplies of
nutriment from the fronds of algæ, from decaying or
dead organic matter, as well as by attacking, from the
outside, some or other of the invertebrate animals.
This, no doubt, is the case with the somewhat similarly constituted Ostracoda, *Paradoxostoma* and its
allies.

The number of joints composing the anterior antenna of what I suppose to be the male of this species
—eleven, as opposed to nine in the female—excites a
doubt as to whether the two forms are specifically the
same. Possibly they are not, but I have, at present,
no means of deciding the question. In other respects
the two are apparently identical.

Genus 4. ACONTIOPHORUS, *Brady, nov. gen.*

(*Solenostoma*, Brady and Robertson.)

Like *Artotrogus*, except that the anterior antennæ are 11-jointed; the posterior, instead of a claw, have two lancet-shaped apical spines, and are 4-jointed; the fifth pair of feet are 2-jointed; the siphon longer than the cephalothorax, and very slender.

1. ACONTIOPHORUS SCUTATUS (*Brady* and *Robertson*). Pl. XC, figs. 1—10.

Solenostoma scutatum, B. & R. On Marine Copepoda taken in the West of Ireland (Ann. and Mag. Nat. Hist., ser. iv, vol. xii, p. 141, 1873).

Body subpyriform (fig. 1); cephalothorax broadly ovate; head united with the first thoracic somite, the segment thus formed being very large and equal to nearly half the entire length of the body; abdomen of the *female* 3-jointed, the first segment large, and composed of two coalescent somites. Posterior angles of all the body-segments rounded off, or only very slightly produced. Anterior antennæ (fig. 2) very short, scarcely one third as long as the first segment of the body, stout at the base, and gradually tapering to the apex, densely clothed on the outer margin and apex with long fine hairs, some of which are plumose; to the seventh joint is attached a long curved olfactory appendage. The following formula represents the relative lengths of the various joints :

1,	2,	3,	4,	5,	6,	7,	8,	9,	10,	11.
9	7	2	1	2	5	2½	2½	3	.3	4

Posterior antennæ (fig. 3) 4-jointed, the last joint armed at the apex with two strong lancet-shaped spines, together with one long and four or five very short setæ; at the base of the external margin are also a few small setæ; the second joint gives origin to a 1-jointed secondary branch, which terminates in a long plumose seta. Mandible (fig. 4) simple, consisting of a short stout peduncle bearing a very long plumose seta (probably also a filiform palp, though I have not seen this). Maxillæ (fig. 5) composed of two stout digits, one of which bears three, the other four, stout, curved, and densely plumose setæ. The foot-jaws are precisely similar to those of *Artotrogus.* Siphon (fig. 6) excessively long and slender, reaching as far as the middle of the caudal segments. Outer and inner branches of the swimming-feet nearly equal in length, 3-jointed, all the joints much constricted at the base (fig. 9), first and second joints dilated at the apex, third elongated and narrow; the distal margins of the first and second joints are strongly dentated, and in the inner branch are, at the outer angles, produced downwards into sharp spines; the marginal spines of the outer branch are long and dagger-shaped, the last joint of both branches bearing a long subulate and much attenuated apical spine. Fifth pair of feet (fig. 10) stout, 2-jointed, first joint shorter than broad, and bearing one long seta, second longer than broad, and furnished with five long, subequal, terminal setæ. Caudal segments about thrice as long as broad, and nearly equal in length to the last two abdominal somites; terminal setæ five, finely plumose, three short and two of moderate length, the longest being more than equal to the length of the abdomen. Length $\frac{1}{26}$th of an inch (·98 mm.).

This, though nowhere an abundant species, is generally distributed round the British Islands; the following are the localities in which it has been observed. Taken in the surface-net at night in Roundstone and Westport Bays, and on *Laminaria saccharina* in Clifden Bay, Ireland; dredged off Portincross, Ayrshire, and in forty fathoms off St. Agnes (Scilly Islands); in thirty to thirty-five fathoms off Robin Hood's Bay, Staiths, and Red Cliff (Yorkshire); and in twenty-seven fathoms off Hawthorn (Durham).

2. ACONTIOPHORUS ARMATUS, *nov. sp.* Pl. LXXXVII, figs. 8—15.

Ascomyzon ornatum, Brady and Robertson. Brit. Assoc. Report, . p. 197 (1875).

Anterior antenna (fig. 8) 6-jointed, very slender, densely clothed with fine setæ, some of which are plumose; the relative lengths of the joints are as follows:

1,	2,	3,	4,	5,	6,	7,	8,	9,	10,	11,	12,	13,	14,	15,	16.
7	2	7	3	1	1	3	2	2	2	2	4	4	1¼	3	3

Some of the setæ of the basal joints are very stout and almost spine-like. Mandible (fig. 10) simple, filiform ; palp consisting of one long simple seta. Maxilla two-lobed (fig. 11), one lobe minute and bisetose, the other larger and furnished with three long plumose setæ. Posterior antennæ and foot-jaws nearly like those of *A. scutatus.* Fifth foot (fig. 15 *a, a*) composed of one broad subovate branch and bearing five long marginal setæ. Abdomen (female) of three segments, the first constituting more than half the length of the abdomen, and composed of two coalescent somites ; second segment scarcely half as long, posterior angles of both produced backwards and overlapping the succeeding segment ; third segment about as long as the second ; caudal appendages scarcely longer than broad, each bearing five plumose setæ, the largest of which are longer than the abdomen. Length $\frac{1}{16}$th of an inch (1·5 mm.).

Two specimens only found, both dredged off the Yorkshire coast, near Scarborough and Robin Hood's Bay.

The word *ornatum* was applied to this species by a misprint in the ' British Association Report,' and is not properly applicable ; *armatum* was chosen for the specific name with reference to the spinous setæ of the anterior antennæ.

NOTE ON THE GENUS *TEMORA*.

In the definition of this genus (vol. i, p. 53) the inner branches of the first, second, third, and fourth pairs of feet are stated to be two-jointed. This is correct as far as the typical species, *T. longicornis*, is concerned; but, in the case of *T. velox*, the inner branch of the *first* pair consists of only *one* joint. It is therefore necessary to amend the generic character as follows :—Inner branch of the first pair of feet one- or two-jointed; of the second, third, and fourth pairs two-jointed.

Dr. Claus makes the one-jointed inner branch of the first pair a generic character, and it would therefore appear that he has found it to hold good in *T. longicornis*, on which species and *T. armata* his description is founded. But, though I have noted in my remarks on *T. longicornis* that there is often, in British specimens, only an imperfect division of the branch into two joints, there can be no doubt that this is a sign of immaturity or imperfect development. Possibly there may be race-variations, and it would be interesting to know whether the specimens described by Dr. Claus— probably from Heligoland—exhibit constant characters in the jointing of the first foot.

INDEX

TO

VOLUMES I, II, AND III.

PRINTED BY J. E. ADLARD, BARTHOLOMEW CLOSE.

PLATE LXXXIII.

Cyclopicera gracilicauda.

Fig. 1. Anterior antenna.
    2. Posterior antenna.
    3. Mandible.
    4. Maxilla.
    5. Labrum.
    6. Anterior foot-jaw.
    7. Posterior foot-jaw.
    8. Foot of first pair.
    9. Foot of fourth pair.
    10. Abdomen ; *a,* fifth foot.

Corycœus anglicus.

    11. Adult male.
    12. Anterior antenna.
    13. Posterior antenna of female.
    14. Posterior antenna of male.
    15. Tail of male.

Plate 83.

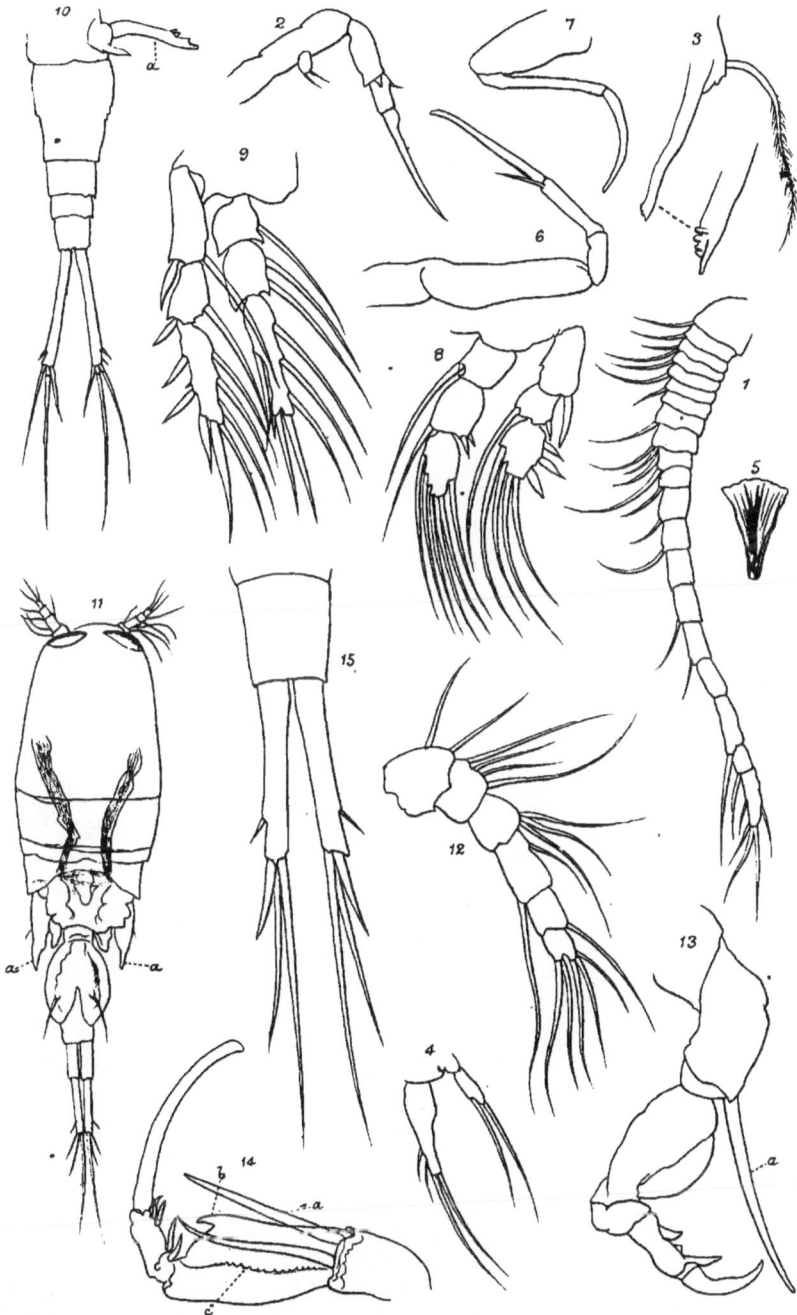

G.S.Brady del.
A.T.Hollick lith.

W.West & Co. imp.

1—10, Cyclopicera gracilicauda.
11—15, Corycoeus anglicus.

PLATE LXXXIV.

Cylindropsyllus lœvis.*

Fig. 1. Adult female.
    2. Anterior antenna.
    3. Posterior antenna.
    4. Posterior foot-jaw (?).
    5. Foot of second pair.
    6. Foot of fourth pair.
    7. Foot of fifth pair.
    8. Caudal lamina with setæ.

Corycœus furcifer.

    9. Mouth organs; *a*, labrum; *b*, mandible; *c*, maxilla; *d*, anterior foot-jaw.

Corycœus anglicus.

   10. *a*, mandible; *b*, palp; *c*, maxilla.
   11. Anterior foot-jaw.
   12. Foot of third pair.
   13. Foot of fourth pair.
   14. Lower thoracic and first abdominal segments.

* In the inscription of the Plate, for *Cylindrosoma* read *Cylindropsyllus.*

1—8, Cylindrosoma lœvis.
9. Corycœus furcifer.
10–14   „   anglicus.

G.S.Brady del.
A.T.Hollick lith.

W.West & Co. imp.

PLATE LXXXV.

Lichomolgus fucicolus.

Fig. 1. Adult male.
 2. Anterior antenna.
 3. Posterior antenna.
 4. *a*, mandible; *b*, maxilla.
 5. Anterior foot-jaw.
 6. Posterior foot-jaw of female.
 7. Posterior foot-jaw of male.
 8. Foot of first pair.
 9. Foot of fourth pair.
 10. Abdomen of female ; *a*, fifth foot.
 11. Abdomen of male.

Lichomolgus forficula. *

 12.† Mandible.
 13. Posterior foot-jaw of male.
 14. Inner branch, fourth foot.
 15. Foot of fifth pair.
 16. Abdomen of male.

* In the Plate, for *forcicula* read *forficula*.
† In the Plate, *for* 12—18 *read* 12—16.

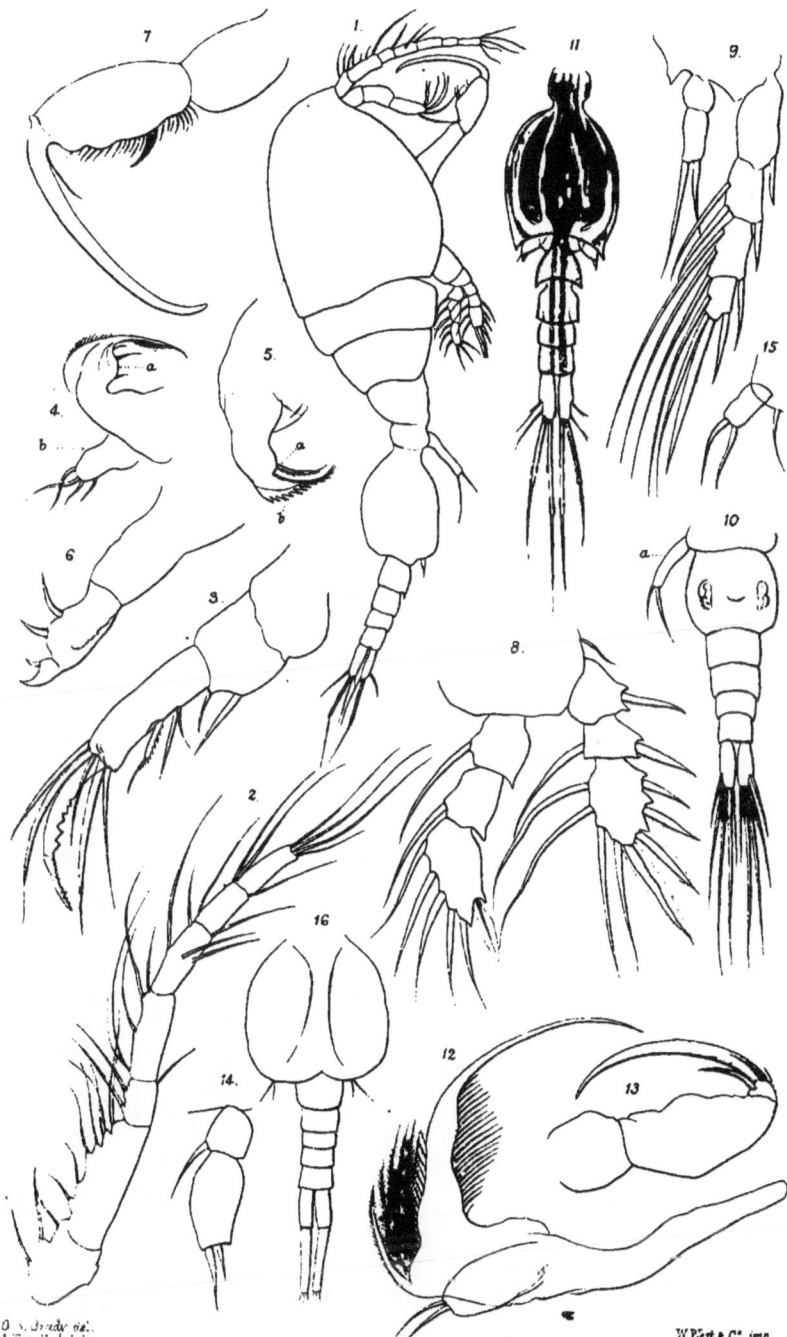

1–11 Lichomolgus fucicolus.
12–16   „   forcifula.

PLATE LXXXVI.

Lichomolgus liber.

Fig. 1. Adult male.
 2. Anterior antenna.
 3. Posterior antenna.
 4. Mandible.
 5. Anterior foot-jaw.
 6. Posterior foot-jaw of female.
 7. Posterior foot-jaw of male.
 8. Foot of first pair.
 9. Foot of second pair.
 10. Foot of fourth pair.
 11. Foot of fifth pair.
 12. Abdomen of female.
 13. Abdomen of male.

Lichomolgus forficula.

 14. Anterior antenna.
 15. Posterior antenna.
 16. Anterior foot-jaw.
 17. Posterior foot-jaw of female.
 18. Abdomen of female.

Plate 86.

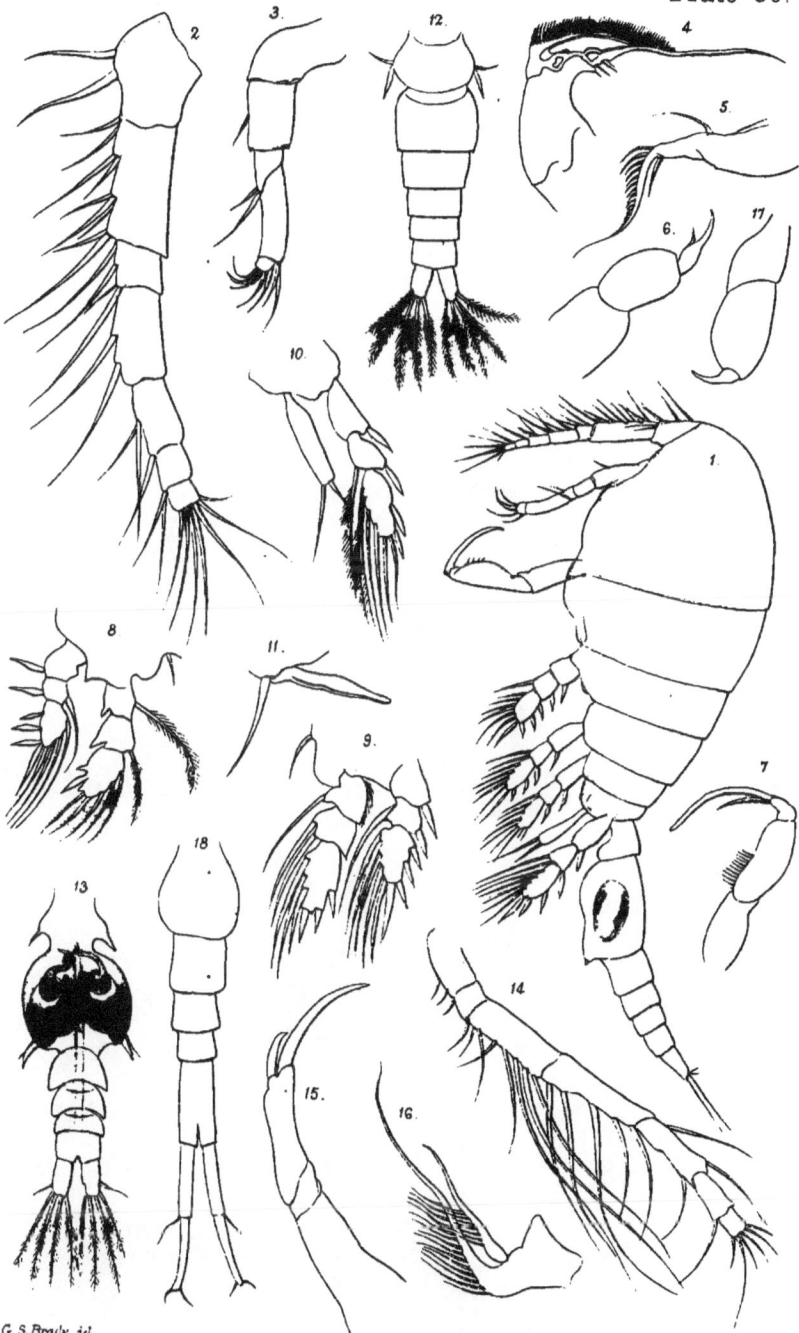

1—13 Lichomolgus liber.
14—18   „   forficula.

PLATE LXXXVII.

Lichomolgus arenicolus.

Fig. 1. Adult male.
2. Anterior antenna, female.
3. Mandible ; *a*, maxilla (?).
4. Anterior foot-jaw.
5. Posterior foot-jaw of male.
6. Foot of first pair.
7. Foot of fifth pair.

*Acontiophorus armatus.**

8. Anterior antenna.
9. Posterior antenna.
10. Mandible.
11. Maxilla.
12. First foot-jaw.
13. Second foot-jaw.
14. Siphon.
15. Abdomen and fifth foot.

* In the inscription of the Plate, for *Solenostoma armatum* read *Acontiophorus armatus.*

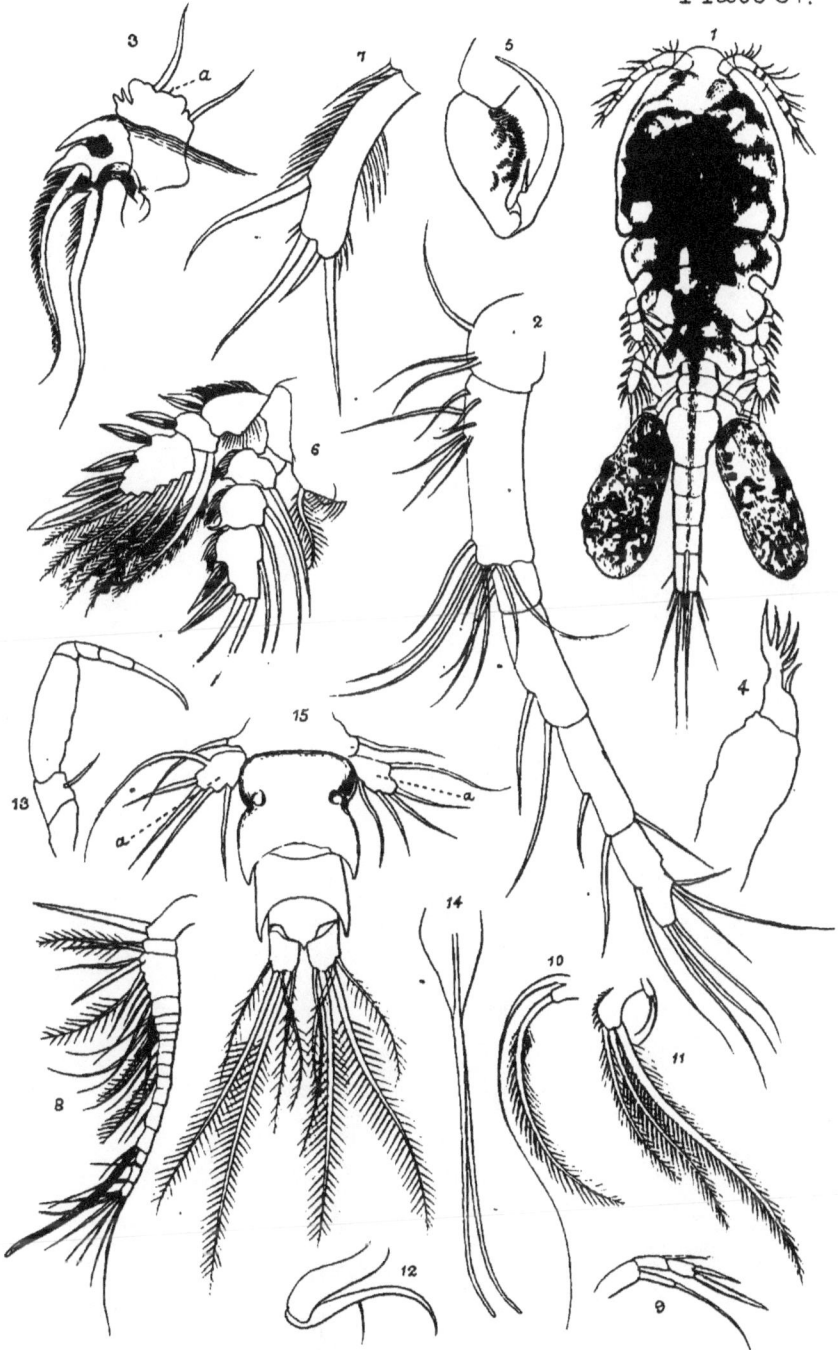

Plate 87.

PLATE LXXXVIII.

Lichomolgus Thorellii.

Fig. 1. Anterior antenna.
    2. Posterior antenna.
    3. Maxilla.
    4. Anterior foot-jaw.
    5. Posterior foot-jaw of female.
    6. Posterior foot-jaw of male.
    7. Foot of fourth pair.
    8. Abdomen and tail of female.
    9. Abdomen and tail of male.

Lichomolgus furcillatus.

    10.* Anterior antenna.
    11. Mandible.
    12. Anterior foot-jaw.
    13. Foot of first pair.
    14. Abdomen and tail ; *a* fifth foot.

* In the inscription of the Plate, *for* figs. 1—14 *read* 10—14.

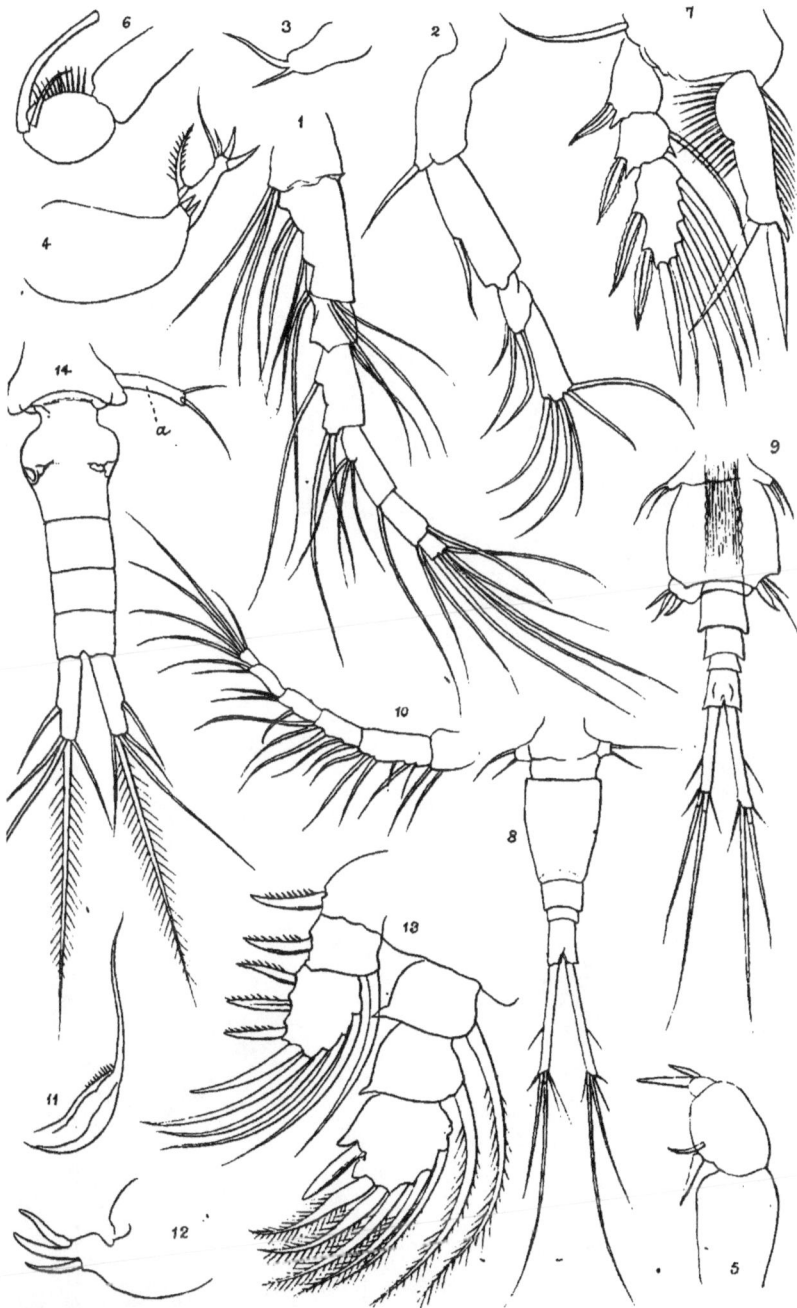

Plate 88.

1—9 Lichomolgus Thorellii.
1—14   „   furcillatus.

PLATE LXXXIX.

Cyclopicera nigripes.

Fig. 1. Female seen from below.
- *a.* Anterior antenna.
- *b.* Posterior antenna.
- *c.* Mandible.
- *d.* Maxilla.
- *e.* Anterior foot-jaw.
- *f.* Posterior foot-jaw.
- *g–k.* 1st—5th pairs of feet.
- *l.* Labrum.

2. Anterior antenna of female.
3. Posterior antenna.
4. Mandible.
5. Maxilla.
6. Labrum.
7. Anterior foot-jaw.
8. Posterior foot-jaw.
9. Foot of second pair.
10. Foot of fifth pair.
11. Abdomen and tail of male.

Cyclopicera lata.

12. Abdomen of female,

Plate 89.

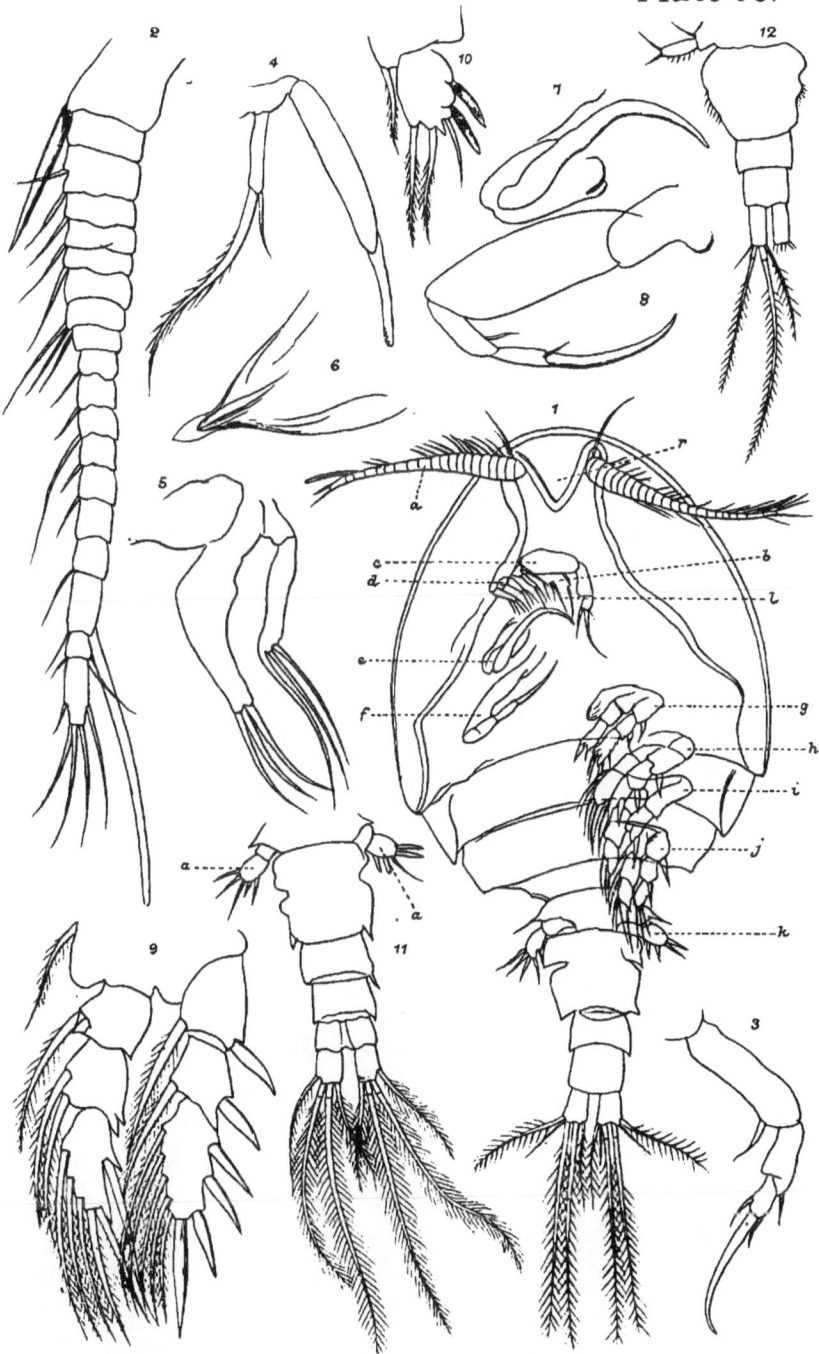

1–11.Cyclopicera nigripes.
12,   ,,   lata.

PLATE XC.

Acontiophorus scutatus* (female).

Fig. 1. Female seen from below,
  a. Anterior antenna.
  b. Posterior antenna.
  c. Mandible.
  d. Maxilla.
  e. Anterior foot-jaw.
  f. Posterior foot-jaw.
  s. Siphon.
  2. Anterior antenna.
  3. Posterior antenna.
  4. Mandible.
  5. Maxilla.
  6. Siphon.
  7. Anterior foot-jaw.
  8. Posterior foot-jaw.
  9. Foot of third pair.
  10. Foot of fifth pair.

Cyclopicera lata.

  11. Anterior antenna of female.
  12. Anterior antenna of male.
  13. Posterior antenna.
  14. Maxilla.

---

* In the inscription of the Plate, for *Solenostoma scutatum* read *Acontiophorus scutatus.*

Plate 90

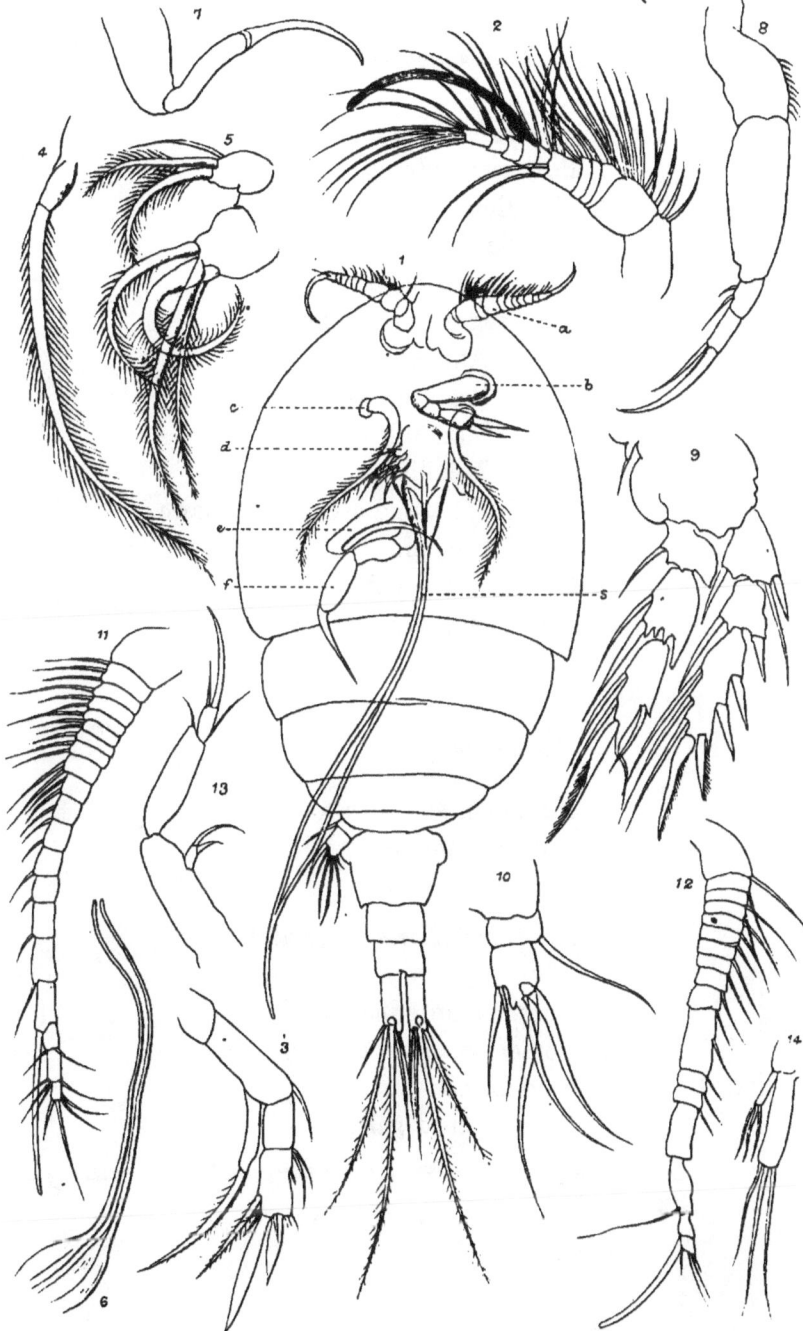

1-10, Solenostoma scutatum♀
11-14, Cyclopicera. lata.

PLATE XCI.

*Artotrogus Boeckii** (female).

Fig. 1. Anterior antenna.
    2. Posterior antenna.
    3. Mandible.
    4. Maxilla.
    5. Posterior foot-jaw.
    6. First segment of body.
       a. Anterior antenna.
       b. Posterior antenna.
      c c. Mandibles.
      d d. Maxillæ.
   e e, f f. First and second foot-jaws.
      g. Siphon.
    7. Foot of third pair.
    8. Abdomen; *a*, fifth foot.
    9. Fifth foot, *var.*

Cyclopina gracilis (male).

   10. Abdomen and fifth foot.
   11. Anterior antenna.

Artotrogus Normani (male).

   12. Anterior antenna.
   13. Posterior antenna.
   14. Second foot-jaw.
   15. Foot of third pair.

* In the inscription of the Plate, for *Artotrogus Lilljeborgii* read *Artotrogus Boeckii*.

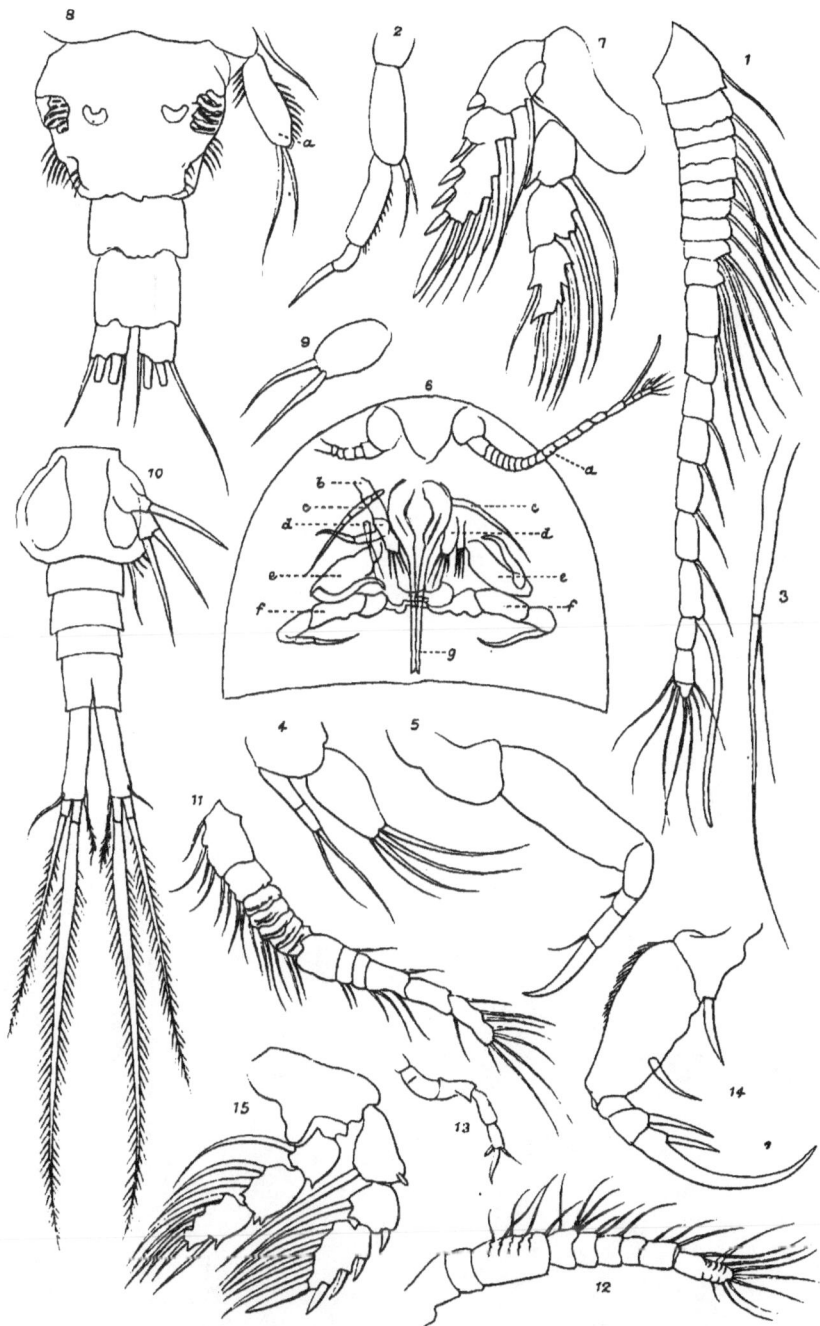

1—9 Artotrogus Lilljeborgii ♀.
10,11 Cyclopina gracilis ♂.
12-15 Artotrogus Normani ♂.

PLATE XCII.

Dyspontius striatus.

Fig. 1. Adult male.
   2. Anterior antenna of female.
   3. Anterior antenna of male.
   4. Posterior antenna.
   5. Maxilla.
   6. Anterior foot-jaw.
   7. Posterior foot-jaw.
   8. Foot of third pair.
   9. Foot of fourth pair.
   10. Appendage of first abdominal segment, male.
   11. Terminal spines of swimming-foot.
   12. Abdomen of female.
   13. Shell-structure.

Artotrogus Normani.

   14. Shell-structure.

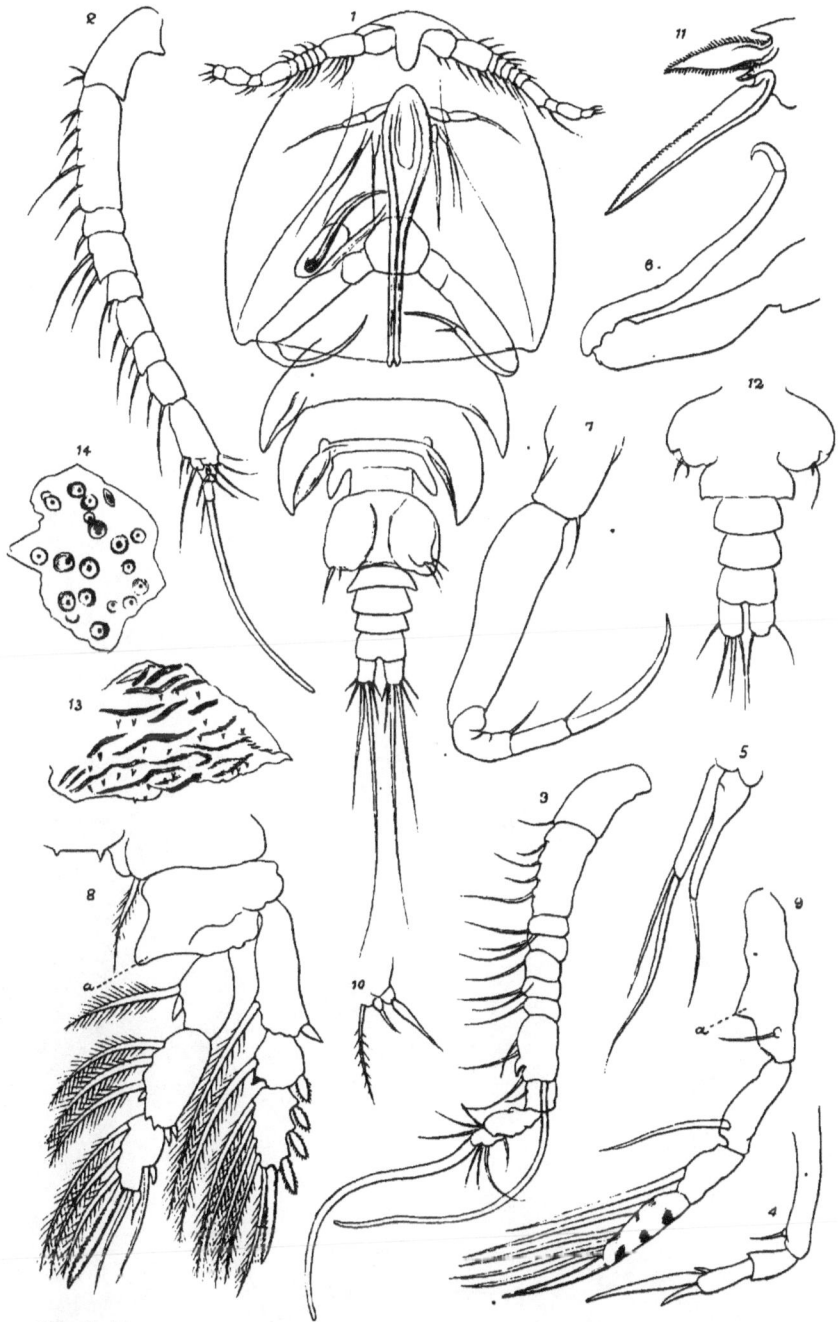

1—13 Dyspontius striatus.
14 Artotrogus Normani.

PLATE XCIII.

Artotrogus magniceps (female).

Fig. 1. Adult female.
    2. Anterior antenna.
    3. Siphon.
        a a. Extremities of siphonal limbs.
        a'. Same more highly magnified.
        b. Mandible.
        b'. Extremity of same more highly magnified.
    4. Maxilla.
    5. First foot-jaw.
    6. Second foot-jaw.
    7. Foot of fourth pair.
    8. Foot of fifth pair.
    9. Shell-structure.

Artotrogus Normani.

   10. Adult male.

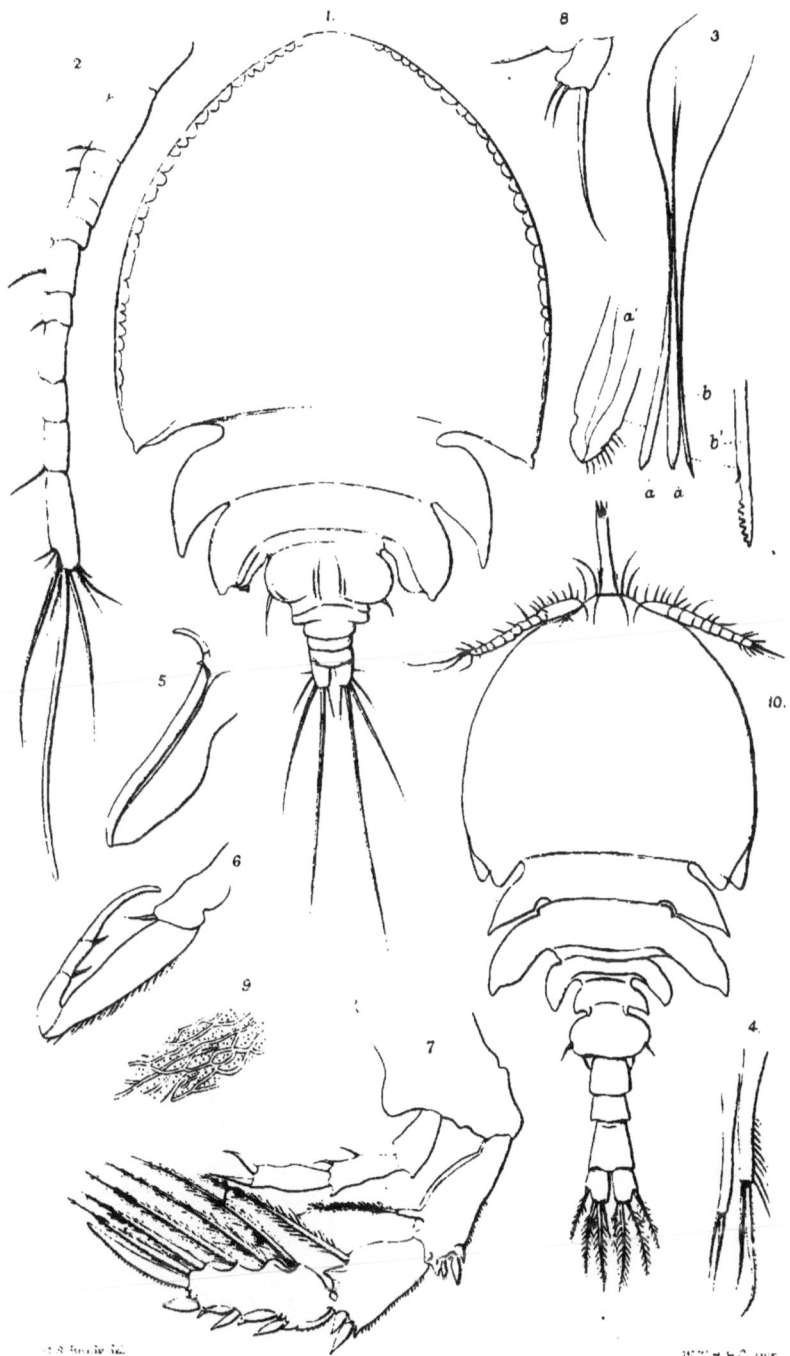

RAY SOCIETY.

INSTITUTED 1844.

FOR THE PUBLICATION OF WORKS ON
NATURAL HISTORY.

ANNUAL SUBSCRIPTION ONE GUINEA.

LIST

OF

COUNCIL, OFFICERS, LOCAL SECRETARIES,
AND SUBSCRIBERS,

TOGETHER WITH THE

TITLES OF THE PUBLICATIONS OF THE SOCIETY,

CORRECTED TO AUGUST, 1880.

Council and Officers of the Ray Society,

Elected 28th May, 1880.

President.

Sir P. DE MALPAS GREY EGERTON, Bart., M.P., F.R.S.

Council.

Sir A. BRADY, F.G.S.
Dr. BRAITHWAITE, F.L.S.
J. DEANE, Esq., F.L.S.
J. FLOWER, Esq., M.A., F.Z.S.
C. H. GATTY, Esq., F.L.S.
F. GRUT, Esq., F.L.S.
E. HARRIS, Esq., F.L.S.
R. HUDSON, Esq., F.R.S.
H. LEE, Esq., F.L.S.
R. M'LACHLAN, Esq., F.R.S.
W. MATHEWS, Esq., F.G.S.

H. T. MENNELL, Esq., F.L.S.
Dr. J. MILLAR, F.L.S.
F. P. PASCOE, Esq., F.L.S.
B. WOODD SMITH, Esq., F.Z.S.
H. T. STAINTON, Esq., F.R.S.
C. STEWART, Esq., F.L.S.
Professor TENNANT, F.G.S.
Capt. C. TYLER, F.L.S.
Dr. E. HART VINEN, F.L.S.
J. J. WEIR, Esq., F.L.S.

Treasurer.

Dr. S. J. A. SALTER, F.R.S., F.L.S., 44, New Broad Street, E.C.

Secretary.

Rev. THOMAS WILTSHIRE, M.A., F.L.S., 25, Granville Park, Lewisham, S.E.

LIST OF LOCAL SECRETARIES.

Aberdeen	Professor Trail.
Bath	J. W. Morris, Esq.
Belfast	Professor Cunningham.
Birmingham	W. R. Hughes, Esq.
Dublin	Dr. W. E. Steele.
Edinburgh	Professor Balfour.
Lancaster	T. Howitt, Esq.
Leeds	L. C. Miall, Esq.
Liverpool	Isaac Byerley, Esq.
Manchester	E. W. Binney, Esq.
Norwich	F. W. Harmer, Esq.
Oxford	Professor Lawson.
Warrington	T. G. Rylands, Esq.

LIST OF SUBSCRIBERS.*

Aberdeen, University of, Aberdeen.
Adlard, J. E., Esq., Bartholomew close, E.C.
Allman, Professor, F.R.S., &c., Sunny hill, Parkstone, Poole, Dorset.
Alvey, Thomas, Esq., 200, Pentonville road, N.
American Institute, New York.
Andrews, Arthur, Esq., Newtown House, Blackrock, Dublin.
Angelin, Professor, Stockholm.
Argyll, Duke of, F.R.S., Argyll Lodge, Kensington, W.
Armstrong, Sir W. G., F.R.S., The Minories, Newcastle-on-Tyne.
Army and Navy Club, 36, Pall Mall, S.W.
Asher, Messrs., 13, Bedford street, W.C.
Ashmolean Society, Oxford.
Asiatic Society, Royal, Bombay.
Athenæum Club, Pall Mall, S.W.
Aubrey, Rev. H. G. W., Rectory, Hale, Salisbury.

Babington, Professor Charles C., M.A., F.R.S., Cambridge.
Babington, the Rev. Professor Churchill, F.L.S., Cockfield Rectory, near Sudbury, Suffolk.
Baer, Herr J., Frankfort.
Baillière, Messrs., 20, King William street, W.C.
Bain, J., Esq., 1, Haymarket, S.W.
Baker, Alfred, Esq., 59, Hagley road, Edgbaston, Birmingham.
Balfour, Professor, M.D., LL.D., F.R.S., L.S., *Local Secretary*, 27, Inverleith row, Edinburgh.
Balfour, F. M., Esq., Trinity College, Cambridge.

* The Subscribers are requested to inform the Secretary of *any errors or omissions* in this List, and of any delay in the transmission of the Yearly Volume.

Balfour, Prof. I. Bailey, D.Sc., 11, Hillhead Gardens, Glasgow.
Baltimore, Peabody Institute.
Bastian, Dr. H. C., F.R.S., F.L.S., 20, Queen Anne street, W.
Bath Microscopical Society, Bath.
Beaufoy, Executors of the late G., Esq., South Lambeth, S.
Belfast Linen Hall Library, Belfast.
Belfast Queen's College, Belfast.
Bell, Dr. W. R., 8, Rutland park villas, Perry hill, Catford bridge,
    S.E.
Bentley, Professor, F.L.S., King's College, Strand, W.C.
Bergen, Museum of, Bergen.
Berlin Royal Library, Berlin.
Binks, J., Esq., Wakefield.
Birmingham Free Library, Birmingham.
Birmingham Old Library, Birmingham.
Birmingham Natural History and Microscopical Society, Birmingham.
Blackwall, John, Esq., F.L.S., Hendre House, near Llanrwst, Denbigh-
    shire.
Blatch, W. G., Esq., Small Heath, Birmingham.
Blomefield, Rev. L., F.Z.S., 19, Belmont, Bath.
Bloomfield, Rev. E. N., M.A., Guestling, near Hastings.
Boston Public Library, U.S., Boston.
Boswell, Dr. J. T., Balmuto, Kirkcaldy, N.B.
Brady, Sir A., F.G.S., Maryland point, Stratford, Essex, E.
Brady, H. B., Esq., F.L.S., Hillfield, Gateshead.
Braikenridge, Rev. G. W., M.A., F.L.S., Clevedon, Bristol.
Braithwaite, Dr. R., F.L.S., The Ferns, Clapham rise, S.W.
Bree, Dr. C. R., F.L.S., East hill, Colchester.
Brevoort, Dr. J. Carson, New York.
Brighton and Sussex Natural History Society, Brighton.
Bristol Microscopical Society, Bristol.
Brockholes, Mrs. J. Fitzherbert, Clifton hill, Garstang, Lancashire.
Brodrick, W., Esq., Little hill, Chudleigh, South Devon.
Brook-Ter., Geo., Esq., Fernbrook, Huddersfield.
Broome, C. E., Esq., M.A., F.L.S., Elmshurst, Batheaston, Bath.
Browell, E. M., Esq., Buckingham Palace, S.W.
Brown, Charles Henry, Esq., London street, Southport, Lancashire.
Browne, Dr. Henry, Woodheys, Heaton Mersey.
Browne, Rev. T. H., F.G.S., High Wycombe, Bucks.
Buckton, G. B., Esq., Weycombe, Haslemere, Sussex.

Burn, Dr. W. B., Ecclesbourne, Bedford hill road, Balham, S.W.
Burton, John, Esq., Lee terrace, Blackheath, S.E.
Bury District Co-operative Provision Society (Limited), Market street, Bury, Lancashire.
Busk, Professor George, F.R.S., F.L.S., 32, Harley street, Cavendish square, W.
Byerley, I., Esq., F.L.S., *Local Secretary*, Seacombe, Cheshire.

Cambridge University Library.
Cambridge, Downing College.
Cambridge, Gonville and Caius College.
Cambridge, St. Catharine's College.
Cambridge, Sidney-Sussex College.
Cambridge, Trinity College.
Campbell, F. M., Esq., Rose hill, Hoddesdon.
Carpenter, Dr. A., High street, Croydon, S.
Carpenter, Dr. W. B., F.R.S., 56, Regent's park road, N.W
Cartwright, Rev. W. H., Butcombe Rectory, Wrington, Somerset.
Carus, Dr. Victor, Leipsic.
Cash, W., Esq., Elmfield terrace, Savile park, Halifax.
Chapman, E., Esq., Frewen Hall, Oxford.
Cheltenham Permanent Library, Cheltenham.
Chicago Library, Chicago.
Christiania, University of.
Christison, Sir R., Bart., 42, Moray place, Edinburgh.
Church, Dr. W. S., 130, Harley Street, W.
Clark, J. A., 11, Duncan place, London fields, Hackney, E.
Clark, J. W., Esq., Scroope House, Trumpington street, Cambridge.
Cleland, Professor, 2, The College, Glasgow.
Clermont, Lord, Ravensdale park, Newry, Ireland.
Collings, Rev. W. T., M.A., F.L.S., Hirzel House, Guernsey.
Colman, Jeremiah J., Esq., M.P., Carrow House, Norwich.
Cooke, Benjamin, Esq., 103, Windsor Road, Southport.
Cooper, Colonel E. H., 42, Portman square, W.
Cooper, Sir Daniel, Bart., 6, De Vere gardens, Kensington Palace, W.
Coppin, John, Esq., Kingfield House, by Corbridge-on-Tyne, R.S.O.
Cork, Queen's College, Cork.
Cornwall, Royal Institution of, Truro.
Craven, Alfred E., Esq., 36, Princes Gate, S.W.

Cresswell, Rev. R., Teignmouth, Devon.
Croft, R. Benyon, Esq., R.N., F.L.S.,, Farnham Hall, Ware, Herts.
Crowley, Philip, Esq., Wadden House, Croydon, S.
Cruickshank, Alexander, Esq., 12, Rose street, Aberdeen.
Cunningham, Professor R. O., *Local Secretary*, Queen's College, Belfast.
Currey, Fred., Esq., F.R.S., Sec. L.S., 3, New square, Lincoln's inn,
    W.C.
Curtis, William, Esq., Alton, Hants.

Darwin, C., Esq., LL.D., F.R.S., Down, Kent.
Dawson, Professor J. W., F.R.S., F.G.S., M'Gill College, Montreal.
Deane, Jas., Esq., F.L.S., 17, The Pavement, Clapham, S.W.
Devon and Exeter Institution, Exeter.
Devonshire, Duke of, 78, Piccadilly, W.
Dickie, Prof. G., M.D., F.L.S., 16, Albyn terrace, Aberdeen.
Dickinson, Wm., Esq., Jun., Warham road, Croydon.
Dickson, Professor Alexander, Glasgow.
Dohrn, Dr. Anton, Naples.
Douglas, J. W., Esq., 8, Beaufort gardens, Lewisham, S.E.
Douglas, Rev. R. C., Manaton Rectory, Moreton Hampstead, Exeter.
Douglas, W. D. R., Esq., Orchardton, Castle Douglas, N.B.
Drewitt, Drewitt O., Esq., Jarrow Hall, Newcastle-on-Tyne.
Drosier, Dr. W. H., Cambridge.
Dublin, National Library.
Dublin, Royal Irish Academy.
Dublin, Trinity College.
Dublin, College of Surgeons.
Dublin, Hon. Society of King's Inn.
Ducie, Earl of, F.R.S., F.G.S., 16, Portman square, W.
Dunning, J. W., Esq., M.A., F.L.S., 24, Old buildings, Lincoln's
    inn, W.C.

East Kent Natural History Society, Canterbury.
Edgeworth, M. P., Esq., F.L.S., 6, Notham gardens, Oxford.
Edinburgh College of Physicians.
Edinburgh, Library of University of.
Edinburgh Museum of Science and Art.
Edinburgh, Royal Society of.

Edinburgh Royal Physical Society.
Edinburgh Royal Medical Society.
Egerton, Sir P. de M. Grey, Bart., M.P., F.R.S., G.S., *President*, 28B,
    Albemarle street, W., and Oulton park, Tarporley, Cheshire.
Elliot, Sir W., F.L.S., Hawick, Roxburgshire.
Elphinstone, II. W., Esq., F.L.S., 2, Stone Buildings, Lincoln's Inn,
    W.C.
England, Royal College of Surgeons of, Lincoln's-inn-fields, W.C.
England, Bank of, Library, London, E.C.
Enniskillen, the Earl of, D.C.L., F.R.S., F.G.S., 65, Eaton place, S.W.
Eyton, T. C., Esq., F.G.S., F.L.S., Eyton, Wellington, Salop.

Ferguson, W., Esq., F.L.S., F.G.S., Kinmundy House, near Mintlaw,
    Aberdeenshire.
Ffarington, Miss M. H., Worden Hall, near Preston.
Fitch, Fred., Esq., F.R.G.S., Hadleigh House, Highbury New Park, N.
Flower, J., Esq., M.A , F.Z.S., Fairfield road, Croydon, S.
Flower, W. II., Esq., F.R.S., Royal College of Surgeons, W.C.
Ford, J., Esq., 1, Market street, Wolverhampton.
Foster, C., Esq., Thorpe, Norwich.
Fowler, Rev. W. W., Repton.
Fox, Rev. W. D., Broadlands, Sandown, Isle of Wight.
Friedlander & Son, Messrs., Berlin.
Fuller, Rev. A., Itchenor, near Chichester.

Galton, Capt. Douglas, F.R.S., F.L.S., 12, Chester street, Grosvenor
    place, S.W.
Garner, Robert, Esq., F.L.S., Stoke-upon-Trent.
Garneys, W., Repton, Burton-on-Trent.
Gatty, C. H., Esq., F.L.S., F.G.S., Felbridge Park, East Grinstead,
    Sussex.
Geological Society, London, W.C.
Geological Survey of India, Calcutta.
George, Frederick, Esq., 10, Finchley road, St. John's wood, N.W.
Gerold and Sons, Messrs., Vienna.
Gibson, G. S., Esq., F.L.S., Saffron Walden, Essex.
Glasgow Philosophical Society, Glasgow.
Glasgow University, Glasgow.

Godman, F. D., Esq., F.L.S., 10, Chandos street, Cavendish square.
Goode, J. F., Esq., 3, Regent place, Birmingham.
Gordon, Rev. George, LL.D., Manse of Birnie, by Elgin, N.B.
Gore, R. T., Esq., 6, Queen square, Bath.
Gottingen, University of, Gottingen.
Gould, John, Esq., F.R.S., L.S., 26, Charlotte street, Bedford square, W.C.
Gray, Wm., Esq., 75, Petergate, York.
Graham, W., Esq., F.R.M.S., Ludgate hill, Birmingham.
Green, R. Y., Esq., Newcastle-on-Tyne.
Grut, Ferdinand, Esq., 9, King street, Southwark, S.E.
Günther, Dr., F.R.S., British Museum, W.C.

Hackney Microscopical and Natural History Society, 194, Mare street, Hackney, E.
Haeckel, Professor, Jena, Prussia.
Hailstone, Edward, Esq., F.S.A., Walton Hall, Wakefield.
Hamburgh Town Library, Hamburgh.
Hamilton, Dr. E., F.L.S., F.G.S., 9, Portugal street, Grosvenor square, W.
Hancock, John, Esq., Newcastle-on-Tyne.
Harford, F., Esq., Ocean Marine Insurance Company, 2, Old Broad street, E.C.
Harmer, F. W., Esq., *Local Secretary*, Oakland House, Cringleford, Norwich.
Harper, P. H., Esq., 30, Cambridge street, Hyde Park, W.
Harris, Edw., Esq., F.G.S., Rydal Villa, Longton Grove, Upper Sydenham.
Harris, Dr. F., F.L.S., 24, Cavendish square, W.
Harvey, Dr. J. R., 7, Upper Merrion street, Dublin.
Harvard College, Cambridge, U.S.
Hawkins, Dr. B. L., Woburn, Beds.
Hayek, Herr Gustav Edler von, Vienna.
Hepburn, Sir T. B., Bart., Smeaton, Preston Kirk, N.B.
Hey, Samuel, Esq., 36, Albion street, Leeds.
Hicks, Dr. John B., F.R.S., 24, George street, Hanover square, W.
Hicks, Dr. J. Sibley, 2, Erskine Street, Liverpool.
Hillier, J. T., Esq., 4, Chapel place, Ramsgate.
Hilton, James, Esq., 60, Montagu square, W.

Hinde, R., Esq., Lancaster.
Hoest, Dr., Copenhagen.
Holdsworth, E. W. H., Esq., F.L.S., 84, Clifton hill, Abbey road, N.W.
Hooker, Sir J., C.B., M.D., F.R.S., Kew, W.
Hope, A. J. B., Esq., M.P., 1, Connaught place, W.
Hopkinson, J., Esq., F.Z.S., F.G.S., Wandsford House, Watford.
Houghton, Rev. W., F.L.S., Preston Rectory, Wellington, Salop.
Hovenden, F., Esq., Glenlea, Thurlow Park, Dulwich, S.E.
Howden, Dr. J. C., Sunnyside, Montrose.
Howitt, Thomas, Esq., F.R.C.S.E., *Local Secretary*, Lancaster.
Huddersfield Literary and Scientific Society.
Huddersfield Naturalists' Society.
Hudson, R., Esq., F.R.S., F.G.S., Clapham common, S.W.
Hughes, W. R., Esq., F.L.S., *Local Secretary*, Thorne Villa, Hands-
    worth, Birmingham.
Hull Subscription Library.
Humphry, Professor, F.R.S., Cambridge.
Hunt, John, Esq., Milton of Campsie, Glasgow.
Hutchinson, R., Esq., 29, Chester street, Edinburgh.
Huxley, Professor T. H., F.R.S., Museum Practical Geology, Jermyn
    street, S.W.

Indian Museum, Calcutta.

Jeffreys, J. Gwyn, Dr. F.R.S., F.G.S., Ware Priory, Herts.
Jenner, Charles, Esq., Easter Duddingsten Lodge, Portobello, Edin-
    burgh.
Jordon, Dr. R. C. R., 35, Harborne road, Edgbaston, Birmingham.

Kenderdine, F., Esq., Morningside, Old Trafford, Manchester.
Kibbler, Dr. R. C., 61, King Edward road, Hackney, E.
Kilmarnock Library, Kilmarnock.
Kitson, J., Esq., Elmete Hall, Leeds.
Knapp, A. J., Esq., Llanfoist House, Clifton, near Bristol.

Lancaster Amicable Book Society, Lancaster.

Lawson, Professor, F.L.S., *Local Secretary*, The Botanic Gardens, Oxford.

Lee, Henry, Esq., F.L.S., F.G.S., 43, Holland street, Blackfriars road, S.E.; and Ethelbert House, Margate.

Leeds Philosopical and Literary Society, Leeds.

Leicester, Alfred, Esq., 9, Adelaide terrace, Waterloo, near Liverpool.

Leicester Free Library, Town Hall, Leicester.

Leipzig, University of, Leipzig.

Lendy, Captain A. F., F.L.S., F.G.S., Sunbury House, Sunbury.

Lewis, H. K., Esq., 136, Gower street, W.C.

Lindsay, Charles, Esq., Ridge Park, Lanark, N.B.

Linnean Society, Burlington House, Piccadilly, W.

Lister, Arthur, Esq., F.L.S., Leytonstone.

Liverpool Athenæum, Liverpool.

Liverpool Royal Institution, Liverpool.

Liverpool Library, Lyceum, Liverpool.

Liverpool Medical Institution, Liverpool.

Liverpool Microscopical Society.

Liverpool Free Library, Liverpool.

Lobley, J. Logan, Esq., F.G.S., New Athenæum Club, Pall Mall, S.W.

London Institution, Finsbury circus, E.C.

London Library, 12, St. James's square, S.W.

Lovén, Professor, Stockholm.

Lubbock, Sir J., Bart., M.P., F.L.S., R.S., 15, Lombard street, E.C

McGill, H. J., Esq., Aldenham Grammar School, Elstree, N.W.

McIntosh, W. C., M.D., F.L.S., Perth County Asylum, Murthly.

M'Lachlan, R., Esq., F.R.S., 39, Limes grove, Lewisham, S.E.

Maclagan, Professor Douglas, M.D., F.R.S.E., 28, Heriot row, Edinburgh.

Madras Government Museum, Madras.

Major, Charles, Esq., Red Lion Wharf, 69, Upper Thames street, E.C.

Manchester Free Public Library, Manchester.

Manchester Literary and Philosophical Society, Manchester.

Manners, Geo., Esq., F.L.S., F.S.A., Dingwall road, Croydon.

Mansell-Pleydall, J., Esq., Whatcombe, Blandford.

Martin, C. M., Esq., St. John's Cottage, Merridale road, Wolverhampton.

Margate Microscopical Society.
Mason, P. B., Esq., Burton-on-Trent.
Mathews, W., Esq., M.A., F.G.S., 15, Waterloo street, Birmingham.
Medlycott, W. C., Esq., Ven House, Sherborne, Dorsetshire.
Meiklejohn, Dr. J. W. S., 16, Notting hill square, W.
Mennell, H. T., Esq., F.L.S., 10, St. Dunstan's buildings, Idol lane,
   E.C.
Microscopical Society, Royal, King's College, Strand, London.
Millar, Dr. John, F.L.S., F.G.S., Bethnall House, Cambridge road, N.E.
Milner, E., Esq., Springfield, Warrington.
Mitchell, P. S., Esq. (Messrs. Barber Brothers) Cowper's court, Corn-
   hill, E.C.
Mitchell Library, the, Glasgow.
Mivart, Prof. St. George J., F.R.S., 71, Seymour street, Hyde park, W.
Morris, J. W., Esq., F.L.S., *Local Secretary*, Belmont, Bath.
Moseley, Sir T., Rolleston Hall, Burton-on-Trent.
Munich Royal Library, Munich.
Murray, J., Esq., 3, Clarendon crescent, Edinburgh.
Museum of Economic Geology, London, S.W.

Naylor, John, Esq., Bank, King street, Liverpool.
Naylor, M. E., Esq., Wakefield.
Nelson, Rev. John, Alborough Rectory, Hanworth, Norwich.
Newcastle Literary and Philosophical Society, Newcastle.
Newman, J. P., Esq., 54, Hatton garden, E.C.
Noble, Capt. Jesmond, Dene House, Newcastle-on-Tyne.
Norfolk and Norwich Literary Institution, Norwich.
Norman, Rev. A. Merle, Burnmoor Rectory, Fencehouses, Durham.
Nottingham Free Library.
Nottingham High School.
Nottingham Literary and Philosophical Society, Nottingham.

Owens College, Manchester.
Oxford Magdalen College.

Paisley Philosophical Society, Paisley.
Parke, Geo. H., Esq., Infield Lodge, Barrow-in-Furness.

Parker, W. K., Esq., F.R.S., 36, Claverton street, S.W.
Pascoe, F. P., Esq., F.L.S., 1, Burlington road, Westbourne Park, W.
Peck, R. Holman, Esq., B.A., F.L.S., Elmfield, Penge lane, S.E.
Peckover, Algernon, Esq., F.L.S., Wisbeach.
Peel Park Library, Salford, Lancashire.
Penny, Rev. C. W., Wellington College, Wokingham.
Penzance Public Library, Penzance.
Phené, J. S., Esq., LL.D., F.S.A., 5, Carlton terrace, Oakley street, S.W.
Philadelphia Academy of Natural Sciences, Philadelphia.
Plymouth Institution, Athenæum, Plymouth.
Power, H., Esq., 37A, Great Cumberland place, Hyde Park, W.
Pumphrey, C., Esq., Southfield, King's Norton, near Birmingham.
Pye-Smith, Dr. P. H., 56, Harley street, Cavendish square.

Quekett Club, University College.

Radcliffe Library, Oxford.
Ramsay, Professor A., F.R.S., Museum of Economic Geology, S.W.
Reader, Thomas, Esq., 39, Paternoster row, E.C.
Rigby, Samuel, Esq., Bruche Hall, near Warrington.
Ripon, Marquis of, F.R.S., F.L.S., 1, Carlton gardens, S.W.
Robinson, Rev. George, Tartaraghan, Loughgall, Armagh.
Rolleston, Professor, M.D., F.R.S., Oxford.
Roper, F. C. S., Esq., F.L.S., F.G.S., Palgrave House, Eastbourne.
Rothery, H. C., Esq., M.A., F.L.S,, 94, Gloucester terrace, Hyde Park, W.
Royal Institution, Albemarle street, W.
Royal Medical and Chirurgical Society, 53, Berners street, W.
Royal Society, Burlington House, London, W.
Rowe, J. B., 16, Lockyer street, Plymouth.
Rylands, T. G., Esq., F.L.S., *Local Secretary*, High Fields, Thelwall,
    near Warrington.

Salter, Dr. S. J. A., F.R.S., *Treasurer*, 44, New Broad street, E.C.
Salvin, Osbert, Esq., F.L.S., 10, Chandos street, Cavendish square.
Sanders, Alfred, Esq., F.L.S., Milton, Sittingbourne, Kent.
Sanford, W. A., Esq., F.G.S., Nynehead Court, near Wellington,
    Somersetshire.

Scientific Club, 7, Savile row, W.
Sclater, P. L., Esq., M.A., Ph.D., F.L.S., R.S., 11, Hanover square, W.
Scott, Dr. Wm., Lissenderry, Aughuacloy, Ireland.
Sharp, I., Esq., F.G.S., Culverden hill, Tunbridge Wells.
Sheffield Literary and Philosophical Society, Sheffield.
Sion College Library, London Wall, E.C.
Slack, H. I., Esq., F.G.S., Ashdown Cottage, Forest row, Sussex.
Sladen, Rev. E. H. M., The Gore, Bournemouth.
Slatter, Rev. John, The Vicarage, Streatley, Reading.
Slatter, F. J., Esq., Evesham.
Sloper, G. E., Esq., Devizes.
Smart, Robert B., Esq., 176, Waterloo place, Oxford road, Manchester.
Smith, Basil Woodd, Esq., F.R.A.S., Branch hill, Hampstead, N.W.
Smith, Capt. R., Frankfort Avenue, Rathgar, Dublin.
Somersetshire Archæological and Natural History Society, Taunton.
Sutheran, Messrs., 136, Strand, W.C.
South London Microscopical Club.
Southport Free Library.
Spicer, Messrs., Brothers, 19, New Bridge street, Blackfriars, E.C.
St. Andrew's University Library, St. Andrew's.
Stainton, H. T., Esq., F.R.S., L.S., Mountsfield, Lewisham, S.E.
Stebbing, Rev. T. R. R., Warberry House, Bishopsdown Park, Tunbridge Wells.
Steele, Dr. W. E., *Local Secretary*, 15, Hatch street, Dublin.
Stephenson, J. W., Esq., Equitable Assurance Office, Mansion-house street, E.C.
Stewart, C., Esq., F.L.S., St. Thomas's Hospital, Newington, S.W.
Stockholm Royal Academy, Stockholm.
Stoke Newington Medical Society, per R. Harris, Esq., Secretary, 57, Darnley road, Hackney.
Stowell, Rev. H. A., Breadsall Rectory, near Derby.
Strasbourgh University Library.
Stroud Natural History and Philosophical Society.
Stubbins, J. Esq., Chester College, Old lane, Halifax.
Sunderland Subscription Library, Sunderland.
Swain, E., Esq., 34, Elsham road, Kensington, N.
Swanston, W., Esq., F.G.S., 50, King street, Belfast.

Tebbs, H. V., Esq., 15, Knight Rider street, Doctors' Commons, E.C.

Tennant, Professor James, F.G.S., 149, Strand, W.C.
Thomson, Professor Allen, Glasgow.
Thomson, Dr. Thomas, F.R.S., 6, Bower terrace, Maidstone.
Thomson, Prof. Sir Wyville, F.R.S., Edinburgh.
Toronto, University of, Canada.
Torquay Natural History Society, Torquay.
Townsend, F., Esq., M.A., Honington Hall, Shipston-on-Stour.
Trail, Prof. W. H., M.B., *Local Secretary*, King's College, Old Aberdeen.
Tristram, Rev. Canon H. B., LL.D., F.R.S., The College, Durham.
Trubner & Co., Messrs., London.
Tudor, Richard A., Esq., 21, Church View, Bootle, Liverpool.
Turner, Professor W., F.R.S.E., Anatomical Museum, University of Edinburgh.
Tyler, Captain Charles, F.L.S., F.G.S., 317, Holloway road, Holloway, N.

Upsala, University of, Sweden.

Varenne, E. G., Esq., Kelvedon, Essex.
Vass, M., Leipzig.
Vicars, John, Esq., sen., Seel street, Liverpool.
Vicary, William, Esq., The Priory, Colleton crescent, Exeter.
Vinen, Dr. E. Hart, F.L.S., 17, Chepstow villas West, Bayswater, W.

Wakefield Mechanics' Institution, Wakefield.
Walker, Alfred O., Esq., Chester.
Warden, Dr. Charles, 272, Hagley road, Edgbaston, Birmingham.
Warrington Museum and Library, Warrington.
Warwickshire Natural History Society, Warwick.
Washington Library of Congress, U.S.
Watkinson Library, Harford, Con., U.S.A.
Watford Natural History Society.
Webster, W., Esq., The Downs, Bideford, Devon.
Weir, J. J., Esq., 6, Haddo villas, Blackheath, S.E.
Wells, J. R., Esq., 20, Fitzroy street, Fitzroy square, W.C.
West Kent Natural History Society, Lewisham, S.E.

White, A., Esq., F.L.S., West Drayton.
White, Dr. F. B., 2, Athol place, Perth.
Wills, A. W., Esq., F.C.S., Wylde Green, Erdington, Birmingham.
Wilson, Dr. E., Westal, Cheltenham.
Wiltshire, Rev. T., M.A., F.L.S., G.S., *Secretary*, 25, Granville park, Lewisham, London, S.E.
Wollaston, G. H., Esq., 117, Pembroke road, Clifton, near Bristol.
Wood, E., Esq., Richmond, Yorkshire.
Woodd, B. T., Esq., Conyngham Hall, Knaresborough, Yorkshire.
Wright, Professor E. P., Trinity College, Dublin.

Yale College, New Haven, U.S.
Yeoman, T. P., Esq., 4, St. Hildas terrace, Whitby.
York Philosophical Society, York.
Young, Dr. J., College, Glasgow.

Zoological Society, 11, Hanover square, W.

LIST OF THE ANNUAL VOLUMES

OF THE

RAY SOCIETY.

FROM THEIR COMMENCEMENT, IN 1844, TO
AUGUST, 1880.

LIST OF THE ANNUAL VOLUMES ISSUED BY THE RAY SOCIETY.

FOR THE FIRST YEAR, 1844.

I. Reports on the Progress of Zoology and Botany. Translated by H. E. Strickland, Jun., M.A., F.R.S., E. Lankester, M.D., F.R.S., and W. B. Macdonald, B.A. 8vo.

II. Memorials of John Ray: consisting of the Life of John Ray, by Derham; the Biographical Notice of Ray, by Baron Cuvier and M. Dupetit Thouars, in the 'Biographic Universelle;' Life of Ray, by Sir J. E. Smith: the Itineraries of Ray, with Notes, by Messrs. Babington and Yarrell. Edited by E. Lankester, M.D., F.R.S. 8vo.

III. A Monograph of the British Nudibranchiate Mollusca. By Messrs. Alder and Hancock. Part I. Ten Plates. Imp. 4to.

FOR THE SECOND YEAR, 1845.

I. Steenstrup on the Alternation of Generations. Translated from the German, by George Busk, F.R.S. Three Plates. 8vo.

II. A Monograph of the British Nudibranchiate Mollusca. By Messrs. Alder and Hancock. Part II. Thirteen Plates. Imp. 4to.

III. Reports and Papers on Botany, consisting of Translations from the German. Translated by W. B. Macdonald, B.A.; G. Busk, F.R.S.; A. Henfrey, F.R.S.; and J. Hudson, B.M. Seven Plates. 8vo.

For the Third Year, 1846.

I. Meyen's Geography of Plants. Translated from the German by Miss Margaret Johnston. 8vo.

II. Burmeister on the Organization of Trilobites. Translated from the German, and edited by Professors T. Bell and E. Forbes. Six Plates. Imp. 4to.

III. A Monograph of the British Nudibranchiate Mollusca. By Messrs. Alder and Hancock. Part III. Eleven Plates. Imp. 4to.

For the Fourth Year, 1847.

I. Oken's Elements of Physio-philosophy. Translated from the German by Alfred Tulk. 8vo.

II. Reports on the Progress of Zoology. Translated from the German by Messrs. Geo. Busk, A. H. Haliday, and A. Tulk. 8vo.

III. A Synopsis of the British Naked-eyed Pulmograde Medusæ. By Professor E. Forbes, F.R.S. Thirteen Plates. Imp. 4to.

For the Fifth Year, 1848.

I. Bibliographia Zoologiæ et Geologiæ. By Professor Agassiz. Vol. I. 8vo.

II. Letters of John Ray. Edited by E. Lankester, M.D., F.R.S. Two Plates. 8vo.

III. A Monograph of the British Nudibranchiate Mollusca. By Messrs. Alder and Hancock. Part IV. Twelve Plates. Imp. 4to.

For the Sixth Year, 1849.

I. Reports and Papers on Vegetable Physiology and Botanical Geography. Edited by A. Henfrey, F.R.S. Three Plates. 8vo.

II. A Monograph of the British Entomostracous Crustacea. By W. Baird, M.D., F.R.S. Thirty-six Plates. 8vo.

For the Seventh Year, 1850.

I. Bibliographia Zoologiæ et Geologiæ. By Professor Agassiz. Vol. II. 8vo.

II. A Monograph of the British Nudibranchiate Mollusca. By Messrs. Alder and Hancock. Part V. Fifteen Plates. Imp. 4to.

For the Eighth Year, 1851.

I. A Monograph of the British Angiocarpous Lichens. By the Rev. W. A. Leighton, M.A. Thirty Plates. 8vo.

II. A Monograph of the Family Cirripedia. By C. Darwin, M.A., F.R.S. Vol. I. Ten Plates. 8vo.

FOR THE NINTH YEAR, 1852.

I. Bibliographia Zoologiæ et Geologiæ.   By Professor Agassiz. Vol. III.   8vo.

II. A Monograph of the British Nudibranchiate Mollusca. By Messrs. Alder and Hancock.   Part VI.   Twelve Plates.   Imp. 4to.

FOR THE TENTH YEAR, 1853.

I. A Monograph of the Family Cirripedia.   By C. Darwin, M.A., F.R.S.   Vol. II.   Thirty Plates.   8vo.

II. A Volume of Botanical and Physiological Memoirs, including Braun on Rejuvenescence in Nature.   Six Plates. 8vo.

FOR THE ELEVENTH YEAR, 1854.

Bibliographia Zoologiæ et Geologiæ.   By Professor Agassiz. Vol. IV.   8vo.   (Completing the work.)

FOR THE TWELFTH YEAR, 1855.

A Monograph of the British Nudibranchiate Mollusca.   By Messrs. Alder and Hancock.   Part VII.   Nine Plates. Imp. 4to.   (Completing the work.)

FOR THE THIRTEENTH YEAR, 1856.

A Monograph of the British Fresh-water Polyzoa.   By Professor Allman, F.R.S.   Eleven Plates.   Imp. 4to.

FOR THE FOURTEENTH YEAR, 1857.

A Monograph of the Recent Foraminifera of Great Britain. By Professor Williamson, F.R.S. Seven Plates. Imp. 4to.

FOR THE FIFTEENTH YEAR, 1858.

The Oceanic Hydrozoa. By Professor Huxley, F.R.S. Twelve Plates. Imp. 4to.

FOR THE SIXTEENTH YEAR, 1859.

A History of the Spiders of Great Britain and Ireland. By John Blackwall, F.L.S. Part I. Twelve Plates. Imp. 4to.

FOR THE SEVENTEENTH YEAR, 1860.

An Introduction to the Study of the Foraminifera. By W. B. Carpenter, M.D., F.R.S., F.L.S., &c., assisted by W. K. Parker, F.R.S., and T. Rupert Jones, F.G.S. Twenty-two Plates. Imp. 4to.

FOR THE EIGHTEENTH YEAR, 1861.

On the Germination, Development, and Fructification of the Higher Cryptogamia, and on the Fructification of the Coniferæ. By Dr. Wilhelm Hofmeister. Translated by Frederick Currey, M.A., F.R.S., Sec. L.S. Sixty-five Plates. 8vo.

FOR THE NINETEENTH YEAR, 1862.

A History of the Spiders of Great Britain and Ireland. By
John Blackwall, F.L.S. Part II. Seventeen Plates.
Imp. 4to. (Completing the work.)

FOR THE TWENTIETH YEAR, 1863.

The Reptiles of British India. By Albert C. L. G. Günther,
M.D., F.R.S. Twenty-six Plates. Imp. 4to.

FOR THE TWENTY-FIRST YEAR, 1864.

A Monograph of the British Spongiadæ. By J. S. Bowerbank,
LL.D., F.R.S. Vol. I. Thirty-seven Plates. 8vo.

FOR THE TWENTY-SECOND YEAR, 1865.

I. The British Hemiptera Heteroptera. By Messrs. J. W.
Douglas and John Scott. Twenty-one Plates. 8vo.
II. A Monograph of the British Spongiadæ. By J. S. Bower-
bank, LL.D., F.R.S. Vol. II. 8vo.

FOR THE TWENTY-THIRD YEAR, 1866.

I. The Miscellaneous Botanical Works of Robert Brown,
D.C.L., F.R.S. Vol. I, containing Geographico-botanical,
and Structural, and Physiological Memoirs. Edited by
J. J. Bennett, F.R.S. 8vo.

II. Recent Memoirs on the Cetacea. By Professors Eschricht, Reinhardt, and Lilljeborg. Edited by W. H. Flower, F.R.S. Six Plates. Imp. 4to.

III. Nitzch's Pterylography, translated from the German. Edited by P. L. Sclater, M.A., Ph.D., F.R.S. Ten Plates. Imp. 4to.

For the Twenty-fourth Year, 1867.

I. A Monograph on the Structure and Development of the Shoulder-girdle. By W. K. Parker, F.R.S. Thirty Plates. Imp. 4to.

II. The Miscellaneous Botanical Works of Robert Brown, D.C.L., F.R.S. Vol. II. 8vo.

For the Twenty-fifth Year, 1868.

I. Vegetable Teratology. By M. T. Masters, M.D., F.L.S. 8vo.

II. The Miscellaneous Botanical Works of Robert Brown, D.C.L., F.R.S. Vol. III. Thirty-eight Plates. Imp. 4to.

For the Twenty-sixth Year, 1869.

A Monograph of the Gymnoblastic or Tubularian Hydroids. By J. Allman, M.D., F.R.S. Part I. Twelve Plates. Imp. 4to.

For the Twenty-seventh Year, 1870.

A Monograph of the Gymnoblastic or Tubularian Hydroids. By J. Allman, M.D., F.R.S. Part II. Eleven Plates. Imp. 4to.

For the Twenty-eighth Year, 1871.

A Monograph of the Collembola and Thysanura. By Sir J. Lubbock, Bart., M.P., F.R.S. Seventy-eight Plates. 8vo.

For the Twenty-ninth Year, 1872.

A Monograph of the British Annelids. By W. C. M'Intosh, M.D., F.R.S.E. Part I. Ten Plates. Imp. 4to.

For the Thirtieth Year, 1873.

A Monograph of the British Annelids. By W. C. M'Intosh, M.D., F.R.S.E. Part I. continued. Thirteen Plates. Imp. 4to.

For the Thirty-first Year, 1874.

A Monograph of the British Spongiadæ. By J. S. Bowerbank, LL.D., F.R.S. Vol. III. Ninety-two Plates. 8vo.

For the Thirty-second Year, 1875.

A Monograph of the British Aphides. By G. B. Buckton, F.R.S. Vol. I. Forty-two Plates. 8vo.

For the Thirty-third Year, 1876.

A Monograph of the British Copepoda. By G. S. Brady, M.D., F.L.S. Vol. I. Thirty-six Plates. 8vo.

FOR THE THIRTY-FOURTH YEAR, 1877.

A Monograph of the British Aphides. By G. B. Buckton, F.R.S. Vol. II. Fifty Plates. 8vo.

FOR THE THIRTY-FIFTH YEAR, 1878.

A Monograph of the British Copepoda. By G. S. Brady, M.D., F.L.S. Vol. II. Forty-nine Plates. 8vo.

FOR THE THIRTY-SIXTH YEAR, 1879.

I. A Monograph of the British Copepoda. By G. S. Brady, M.D., F.L.S. Vol. III. Eleven Plates. 8vo. (Completing the work.)

www.ingramcontent.com/pod-product-compliance
Lightning Source LLC
Chambersburg PA
CBHW020555270326
41927CB00006B/854